MILITARY RECONNAISSANCE

MILITARY RECONNAISSANCE

The Eyes and Ears of the Army

ALEXANDER STILWELL

CASEMATE

Oxford & Philadelphia

Published in Great Britain and the United States of America in 2021 by
CASEMATE PUBLISHERS
The Old Music Hall, 106–108 Cowley Road, Oxford OX4 1JE, UK
and
1950 Lawrence Road, Havertown, PA 19083, USA

Hardcover Edition: ISBN 978-1-61200-950-6
Digital Edition: ISBN 978-1-61200-951-3

A CIP record for this book is available from the British Library

Printed and bound in the United Kingdom by TJ Books

Typeset by Versatile PreMedia Service (P) Ltd

For a complete list of Casemate titles, please contact:

CASEMATE PUBLISHERS (UK)
Telephone (01865) 241249
Email: casemate-uk@casematepublishers.co.uk
www.casematepublishers.co.uk

CASEMATE PUBLISHERS (US)
Telephone (610) 853-9131
Fax (610) 853-9146
Email: casemate@casematepublishers.com
www.casematepublishers.com

Front cover image: Ministry of Defence.
Back cover image: U.S. Navy.

Contents

'Time spent on reconnaissance is seldom wasted.'

- Arthur Wellesley, Duke of Wellington

Introduction

Reconnaissance has long been a key tool to enable military commanders to obtain a picture of the tactical situation and to make informed decisions. An early example of this can be found in the scouts drawn from the people known as *Sciritae* who were deployed by the Spartans and had a privileged position in their order of battle. The Spartans were so aware of the advantage that their sentinels and scouts gave them in military operations that they went to great lengths to keep them secret. As military tactics, weapons and equipment developed over the centuries, methods of scouting and reconnaissance evolved and adapted, but were never discarded.

From the highly toned Spartan warriors, to the scouts employed by Julius Caesar, through the Middle Ages, to the Napoleonic Wars and the mass warfare of the modern era, this book aims to provide a concise but revealing picture of the art of military scouting and reconnaissance, always remaining true to the spirit of the scout – light on their feet, taking only what they need and returning with, or sending back, information that could make the difference between victory and defeat. The scout cannot expect to see everything; it is up to the intelligence experts and commanders to interpret the information.

Scouting and reconnaissance responsibilities are carried out by several different military units, some of which are dedicated to reconnaissance, and others which include reconnaissance as part of a wider brief. For example, the United States Marine Corps combine two activities in one with their scout/sniper teams, based on the logic that a sniper's covert skills also place them in a good position to gather intelligence. Special forces units tasked with a mission which may include direct action are also well placed for reconnaissance.

In the first chapter, I look at the development of military scouting and reconnaissance by the ancients, focusing particularly on the use of scouting and decisive action by Julius Caesar during the Gallic wars and by Alexander the Great, king of Macedon. It will not take long to appreciate how information gleaned from their intrepid scouts and their own aptitude for decisive action contributed to their success in battle.

During the medieval period, reconnaissance was carried out in various ways, including by light cavalry. The battle of Hastings (1066) was but one example of how early information about the dispositions of the opposing force gave the Normans the edge over the English army which, fighting on its own familiar ground, may have been thought to have the advantage. Richard the Lionheart used light cavalry during the crusades to determine the position of Saladin's armies.

During the American wars (American Revolutionary War 1775–1783; War of 1812), backwoodsmen familiar with hunting techniques and the ways of the native American Indians made excellent scouts for military commanders. Frontiersmen like Daniel Boone (1734–1820), sharpshooters like Timothy Murphy (1751–1818) and Daniel Morgan (1736–1802), founder of Morgan's Riflemen, came to the fore as well as new reconnaissance units such as Rogers' Rangers formed by Major Robert Rogers (1731–1795). This chapter describes the growing appreciation of the value of these mobile forces, which were able to blend into the environment and appear when and where they were least expected.

In the Napoleonic era (1800–1815), large bodies of men marched in columns in the open, often in brightly coloured uniforms, but the value of skirmishers was also appreciated. This chapter looks at the development of units such as the French *Voltigeurs* and *Chasseurs*, and the British light infantry, as well as Napoleon's use of Cavalry Scouts in the 1814 campaign, the final campaign of the War of the Sixth Coalition.

Although Napoleon had his eye on India and hoped to enlist Russian support, changes in loyalties and alliances meant that a 'Great Game' was played out between Britain and Russia as they struggled for influence in Afghanistan. Here, British and Russian army officers, often in disguise, embarked on extended reconnaissance missions to gather information about enemy movements and possible invasion routes in the remote Afghan mountain passes. This story would continue to repeat itself to the present day and unfortunately many of its lessons would remain unlearned.

The British Army entered a sharp learning curve when it confronted the Boers of South Africa (1899–1902). These hardened men used to hunting on the veldt were hard to defeat on ground that they knew like the backs of their hands. During the Matabele wars (1693–1894; 1896–1897), men such as Frederick Russell Burnham, who had learned his trade among the American Indians and the Apache Wars (1849–1886), Robert Baden-Powell and the hunter Frederick Courtney Selous made their mark. Burnham was so highly valued by the British Commander-in-Chief General Roberts that, despite being American, he was made Chief of Scouts of the British Army.

The Germans began the First World War (1914–1918) with better trained and more effective snipers than the British, who had to learn the hard way in order to adapt. Soon, a British sniping school was set up and the arts of reconnaissance and camouflage were honed. Despite the relatively static nature of trench warfare, patrols into No Man's Land at night, and later during the day, often yielded valuable intelligence. Having made themselves useful during the Boer War (1899–1902), the Scottish Highland yeomanry regiment formed in 1900 and known as the Lovat Scouts continued to distinguish themselves with their unique observation and tacking skills in the Great War.

Apart from the scouts that were part of regular military squads and platoons, specialised units were formed during the Second World War (1939–1945), with specific missions to operate behind enemy lines. These included the US 6th Army Special Reconnaissance Unit 'Alamo Scouts', six- or seven-man teams who operated behind enemy lines in the Pacific Theatre; the British Long-Range Desert Group (LRDG), which operated against German and Italian forces in North Africa; the Special Air Service (SAS), which, beginning in North Africa, spawned French, Belgian and other associated units that were parachuted into France in advance of the Allied invasion of Normandy in 1944; the Lovat Scouts and US Army Rangers.

The non-global conflicts of the Cold War era in regions such as Vietnam and Malaysia, underlined the need for mobile forces with an intuitive understanding of terrain and local people. The role of special forces grew in importance while nuclear confrontation remained unthinkable. In Malaysia, Borneo and Vietnam, special forces honed their patrol skills and their ability to harness the knowledge of local populations.

In the Falklands War (2 April–14 June 1982) and the Gulf Wars of the 1990s and the new millennium, long-range reconnaissance patrols (LRRP) and special forces were at a premium to provide precise targeting information and intelligence about enemy movements. This chapter covers military, naval and air force units tasked specifically with reconnaissance and intelligence gathering.

As technological advances continued and some military reconnaissance units were stood down, this book examines the role of the human scout against the background of the proliferation of reconnaissance and surveillance unmanned aerial vehicles (UAVs), and other technology.

Timeline

389–384BC	Publius Flavius Vegetius Renatus writes influential work on military tactics, *De Re Militari*.
350BC	Xenophon writes *Hipparchicus*, or *The Cavalry Commander*, in which he sets out principles for reconnaissance and scouting.
334BC	Battle of the Granicus River. Alexander III of Macedon makes use of reconnaissance prior to the battle.
331BC	Battle of Gaugamela. Alexander the Great defeats the forces of Darius III.
58–50BC	The Gallic Wars. Julius Caesar uses scouts during his conquest of Gaul.
1066	William of Normandy carries out extensive reconnaissance prior to the engagement with the English army at Hastings.
1740	Prussian jaegers formed into a military unit.
1744	Gorham's Rangers formed by John Gorham in Massachusetts as a model ranger force.
1755	Rogers' Rangers formed by Major Robert Rogers, becoming the most famous and successful of the early ranger units.

1800	Experimental Corps of Riflemen created by Colonel Coote Manningham and Lieutenant-General Sir William Stewart, GCB.
1803	95th (Rifle) Regiment of Foot formed.
1807	French *Voltigeurs* allocated an honoured position on the left of the line.
1810	Captain Charles Christie and Lieutenant General Sir Henry Pottinger carry out reconnaissance of Herat, marking the start of 'the Great Game' between the British Empire and the Russian Empire over Afghanistan and neighbouring territories in Central and South Asia.
1813	Napoleon creates the Scouts of the Imperial Guard (2nd and 3rd *Régiments d'Éclaireurs de la Garde Impériale*).
1814	Kit Carson and John C. Frémont cross the Sierra Nevada and reach the Sacramento River.
1861	Congress authorises the employment of Native Americans in the armed forces.
25 June 1876	Battle of the Little Bighorn. General Custer's force is annihilated after he ignores the advice of his scouts.
November 1893	Frederick Russell Burnham and Frederick Courtney Selous reach Bulawayo in the First Matabele War.
December 1893	Shangani Patrol massacred by Matabele warriors before Burnham can return with reinforcements.
October 1899–May 1900	Siege of Mafeking. Boys are recruited to run errands during the siege and used as an example by Robert Baden-Powell when he later wrote his book *Scouting for Boys*.

1900	The Lovat Scouts formed by Simon Fraser, 14th Lord Lovat, as a Scottish Highland yeomanry regiment of the British Army. It is commanded in South Africa by Frederick Russell Burnham.
1916	Britain's First Army School of Scouting, Observation and Sniping is established.
1924	Heinz Guderian begins to study the potential use of tanks in reconnaissance operations with cavalry.
July 1940	The Long-Range Desert Group (LRDG) is formed to carry out long-range reconnaissance and direct action behind enemy lines in North Africa.
July 1941	The Special Air Service (SAS) is formed by Lieutenant Colonel Sir Archibald David Stirling, DSO, OBE to carry out raids and reconnaissance behind enemy lines.
September 1942	US Marine Corps form a Scout and Sniper detachment.
November 1943	US 6th Army Special Reconnaissance Unit 'Alamo Scouts' is formed.
1943	Operation *Jedburgh*. The British Special Operations Executive (SOE), Special Air Service (SAS), and the US Office of Strategic Services (OSS), along with organisations like the Free French, parachute mixed teams into France, the Netherlands and the Pacific theatre of operations.
January 1944	Military Assistance Command, Vietnam – Studies and Observation Group (MACV-SOG) set up to carry out strategic reconnaissance and covert action in Vietnam.

17–25 September 1944	Operation *Market Garden*. Both the British 1st Airborne Reconnaissance Squadron and the Reconnaissance battalion of the 9th SS Panzer Division '*Hohenstaufen*' are heavily engaged during the battle.
16–17 December 1944	Battle of Lanzerath Ridge. The Intelligence and Reconnaissance platoon of 394th Regiment, 99th Infantry Division hold off the German 1st Panzer Division and 500 German *Fallschirmjäger* at the beginning of the battle of the Bulge.
June 1948–July 1960	The British Special Air Service, along with Australian and New Zealand counterparts are involved in long-range counter-insurgency patrols and reconnaissance in Malaysia.
July 1964–July 1966	Operation *Claret*. British and Commonwealth forces carry out counterinsurgency and reconnaissance in Borneo during confrontation with Indonesia.
1979	International Long-Range Reconnaissance Patrol School (ILRRPS) established in West Germany.
1991	UK Special Air Service (SAS) and US 1st Special Forces Operational Detachment-Delta (SFOD-D) carry out reconnaissance to locate and destroy Iraqi Scud ballistic missiles.
2001	ILRRPS becomes International Special Training Centre (ISTC).

CHAPTER 1

Ancient Warfare

Warfare in the ancient world has many examples of the development of scouting practices and reconnaissance as commanders recognised the value of gathering advance information about enemy movements and acting upon it. In 389 BC, at the battle of the Elleporus, Dionysius I made good use of scouts whereas his opponent, Heloris of the Italiote League, was not aware that Dionysius's army was only five miles away.

Scouts, Sentinels and Spies in the Greek, Macedonian and Roman Armies

Kataskopoi was a Greek term that loosely covered the work of *exploratores*, *procursatores* and *speculatores*. *Kataskopoi* often worked as spies in Classical Greece, unarmed, in plain clothes or in disguise. The traditional military scout, however, was likely to be lightly armed and wearing a light uniform, in other words without armour, the point being that his mission was to gather information and relay it to the commander rather than engage with the enemy and risk either death or capture.

Libanius refers to *kataskopoi* when he describes their use by Constantius when he was engaging the Sasanian Persians at Singara.

The *kataskopoi* identified the points at which the Persian army were crossing the river Tigris. This information was then reported back to Constantius, giving him time to organise a trap into which the Persians were lured. In this case, therefore, Libanius was describing the work of scouts who might alternatively be called *exploratores*. *Skopoi* was another term for any agents who carried out reconnaissance. They might also carry out surveillance.

Various groups known variously as *exploratores, procursatores* and *speculatores*, formed similar intelligence-gathering tasks and sometimes their roles overlapped. In general, *exploratores* were traditional medium- to long-range scouts who would operate well in advance of the cavalry screen, and at distances from the main force of between 12 and 35 kilometres. One needs to bear in mind that with the means of transport and communications available at the time, distances were fairly modest compared with modern standards. The *exploratores* would provide the long-range picture, while the *procursatores* would work at closer range. The role of the *speculatores* was to provide covert intelligence on matters such as the quality of enemy troops and their morale in a manner similar to modern covert and intelligence forces. Vegetius states in *De Re Militari* that:

> It is necessary, to be well acquainted whether the enemy usually make their attempts in the night, at break of day or in the hours of refreshment or rest; and by knowledge of their customs to guard against what we find their general practice. ... Our spies should be constantly abroad; we should spare no pains on tampering with their men and give all manner of encouragement to deserters. By these means we may get intelligence of present or future designs.

The *exploratores* originated in the cavalry and were often hand-picked men. Generals would sometimes use men from their scouting parties as their personal bodyguards, an indication of the quality of the men who were chosen for these roles.

A general would expect to march his force on ground that had been carefully selected by his *exploratores* as the optimum route to the destination with least danger of ambush, or of being delayed by natural obstacles. As Vegetius states:

> A general, therefore, cannot be too careful and diligent in taking necessary precautions to prevent a surprise on the march and in making proper dispositions to repulse the enemy, in case of such an accident, without loss.
>
> In the first place, he should have an exact description of the country that is the seat of war, in which the distances of places specified by the number of miles, the nature of the roads, the shortest routes, by-roads, mountains and rivers, should be correctly inserted.

A general gains intelligence on the march in a constant flow and from more than one source, as the situation constantly changes, both as the army proceeds, and as the enemy may adjust its position. He goes on:

> A general should also inform himself of all these particulars from persons of sense and reputation well acquainted with the country by examining them separately at first, and then comparing their accounts to come to the truth with certainty.

When Gnaeus Julius Agricola came to the rescue of the beleaguered Legio IX Hispana at Mons Graupius in 83 AD his success was largely due to accurate intelligence from *exploratores* on the line of march of the Caledonian force, which enabled him to plot his own route to reach the battlefield and intervene in good time.

Reconnaissance Tactics of Alexander the Great

Battle of the Granicus River 334 BC

In the first battle of the Persian campaign (334–330 BC) by Alexander the Great, his intelligent use of scouts and reconnaissance is evident, and they were central to providing him with the decisive edge through speed of action based on advanced information about the enemy's dispositions.

Alexander's campaign against the Persians was by way of revenge for the Persian invasions of Greece in 492 BC by Darius the Great and 480 BC by Xerxes I. Alexander's overall strategy was based on speed and surprise and his tactics followed suit.

The historian Arrian of Nicomedia tells us that Alexander marched from Troy to Arisbe and moved on to cross the Hellespont, before marching towards the Persian city of Dascylium. As he marched, Alexander sent scouts out ahead of him, under the command of his officer Amyntas, a squadron of elite Companion cavalry, under the command of Socrates, and four squadrons of advanced scouts. The *Psiloi* were the lightly armed troops that often made up reconnaissance teams. They operated over rough ground independently or were used in conjunction with cavalry as screening forces in front of the advancing army. Alexander also used lightly armed Thracian peltasts, sending them to scout the mountainous area around the Persian Gates.

Aware of Alexander's approach, the Persian provincial governors began to assemble their forces near the town of Zelea. The prominent Greek mercenary Memnon of Rhodes recommended that the Persians should use scorched-earth tactics to slow down the advancing Macedonians but the Persian governors were suspicious of the foreigner's motives and chose instead to advance to the Granicus river, which they hoped would have the effect of breaking up the customary Macedonian close formation and complicate their battle order.

As Alexander advanced in battle order towards the river with his infantry massed in two major formations with cavalry on the wings, he sent out ahead of him a reconnaissance party under the command of Macedonian general Hegelochus. This party consisted of lancers and about 500 light troops. It was this reconnaissance party that spotted the Persian army deployed on the far bank of the Granicus river, and scouts were immediately sent galloping back to warn Alexander.

Before he was even within sight of the enemy, Alexander was able to make his plan of attack and start issuing orders. Once he reached the bank of the river, having had the advantage of considering his moves beforehand, Alexander ordered the cavalry on the left flank of his army, under his trusted bodyguard and commander Ptolemy I Soter, to make a feint attack, which had the desired effect of making the Persians send reinforcements to that area. Ptolemy's attack was driven back, as Alexander had foreseen, and he then ordered Amyntas and Socrates, who had led the initial reconnaissance, to lead their units across the river as the spearhead of the attack. They advanced diagonally across the river to counter the current. On the other side, they were met with a hail of missiles, including lances from the Persian defenders, who dominated the banks and low ground on the other side. Alexander and his retinue soon followed. Although there were many casualties among the Macedonians, the sheer weight of their attack and their determination gradually overcame the Persian resistance.

Once Alexander and his army had established a foothold on the far bank, they made straight for the Persian line. When the

Alexander the Great and Bucephalus in battle with the forces of Darius III. (Wikimedia Commons)

Persian cavalry attacked, Alexander was lucky not to lose his own life. Stunned by a blow from a Persian nobleman, he was almost finished off by another but was rescued just in time by one of his own elite companions, an officer called Cleitus the Black.

As the Macedonian cavalry pursued the Persian cavalry to the left, Macedonian infantry marched through the space to engage the Persian infantry, after which they attacked the Greek mercenaries. Seeing the centre collapse, the Persian cavalry retreated, leaving the infantry to be cut down by the now unstoppable Macedonian advance.

Once Alexander had attacked various towns and passed the Cilician Gates, the Persian Emperor Darius III came up behind Alexander's army to surprise him.

Alexander's first decision was to send a reconnaissance party to check that the way was clear for him to move back towards the Persian army. First, his army advanced along a narrow coast road and then, when they reached open ground, his army was arranged in battle order, keeping his left flank close to the sea to avoid being outflanked by the Persians.

On Alexander's right wing, he sent forward his advanced scouts under the command of the Macedonian General Protomachus, commander of the *prodromoi* light cavalry, and the Paeonian light cavalry under the command of Ariston of Paionia, along with the archers under the command of Antiochus I Soter, to probe the hills. These advanced troops forced the enemy to retreat from their positions and move further up the hills. Alexander then personally led an attack on the right wing, causing the enemy to fall back. The successful attack on the right wing enabled the Macedonians to move left towards the centre, where a fierce battle was being waged with Greek mercenaries under Persian command. Soon, the Persian line broke and a rout followed, which included Darius himself, who was lucky to get away, first on a chariot and then, stripped of his regal clothing, on a horse.

Although, in this case, Alexander's scouts were deployed not so much for reconnaissance but as a probing force, their success and overall contribution to the victory underlines the effectiveness of light, elite reconnaissance forces, and is an example of reconnaissance in force.

After Alexander crossed the Tigris river, he continued to send out his scouts to try to pinpoint the location of Darius's forces. Four days after crossing the river, scouts returned to report that they had spotted Persian cavalry. Alexander immediately disposed his army in battle order. When more of his scouts rode in to give him more details of the enemy's strength (an estimated 1,000 cavalry), Alexander rode with his Paeonian rangers, the Royal Squadron and a squadron of Companions to make a reconnaissance in force, while the rest of the army followed behind. Alexander's elite mounted force rode down the Persian unit, killing some, capturing others and causing the rest to flee. The captives informed Alexander that a much larger Persian force under Darius was not far away at a place called Gaugamela. This information enabled Alexander to rest his army and make preparations.

After about four days, Alexander ordered his army forward under cover of night with the intention of attacking the enemy at dawn. Once they had passed over the crest of a ridge, Alexander saw the Persian army in battle order about four miles away. Rather than maintaining the momentum and mounting an immediate attack, as he had done so often before, Alexander took the advice of one of his generals, Parmenion, to carry out a reconnaissance of the ground, in case there were any hidden traps or obstacles, and to check the enemy's dispositions.

Alexander then went out personally with his light infantry and Companion cavalry to conduct a thorough reconnaissance of the whole area, taking into account the terrain on which the battle would be fought, and likely enemy movements. Only when this was complete did he return to gather his commanders for a final briefing. On this occasion, not only had Alexander's reconnaissance confirmed that there were no major defence works, but also that the delay was forcing the Persians to remain at the ready. The fear of an imminent attack was also sapping their morale.

In his book, *Hipparchicus*, Xenophon recommends that generals should carry out personal reconnaissance when possible. Alexander was an example of this, as was Xenophon himself. When they could not go in person, generals should send only highly trusted men.

As a result of his reconnaissance and knowledge of the ground, when Alexander ordered his army to advance, he moved it to the right, away from the area he knew had been cleared by the Persians, so that their bladed chariots could operate more effectively. Although some chariots did attack, they were soon disarmed or overwhelmed by the Macedonians. The battle raged on and at length the Macedonians achieved victory. Alexander also captured all of Darius's treasure, though he was not able to capture the king himself.

The Persian Gates

Repeatedly in the Persian campaign, we see Alexander moving swiftly and decisively. Almost invariably on these fast-moving expeditions, he took with him the *Agrianes,* archers and advanced scouts. The attack on the Persian Gates (330 BC) was just such an example. Seeing that the distinguished satrap or governor of the province, Ariobarzanes the Brave, had placed a considerable force in front of the gates along with physical defences, Alexander decided to outflank them. Following a difficult path, he and his men surprised enemy positions in the hills, before mounting a surprise attack on the main defences at the gate.

Alexander was determined to track down and capture Darius, and his pursuit of the Persian king was relentless. Once again, Alexander moved ahead of the main army, accompanied by his most trusted units, the Companions, advanced scouts and elite light infantry. In this event, before Alexander could catch up with him, Darius was killed by his own escort who realised that there was no longer any chance of escape.

Use of Intelligence

While serving as a senior commander in Thrace in 377 BC, the Roman consul Frigeridus received a warning from his *exploratores* that a Goth army under Farnobius was approaching. Frigeridus withdrew his force to Illyricum and waited for a more suitable time to confront and defeat the Goths.

When faced with the wily king Jugurtha of Numidia (160–104 BC) and having watched the Roman general Metellus fail to bring him to heel, the Roman General Gaius Marius (157–186 BC) realised that he would need to employ more subtle tactics. Having been elected as consul and side-lined Lucius Cecilius Metellus, Marius employed a full range of intelligence activities to subvert Jugurtha and finally capture him with the help of General Lucius Cornelius Sulla (138–78 BC).

Similarly, while the Carthaginian General Hannibal Barca threatened Roman armies in southern Italy after the battle of Lake Trasimene in 217 BC, it was only when Roman generals such as Scipio Africanus (236–183 BC) learned to use intelligence to best effect, that they were able to outwit Rome's most dangerous enemy. After the disastrous defeat at Cannae (216 BC), Roman forces refused to offer battle to Hannibal but instead engaged in a skilful game of manoeuvre that gradually depleted Hannibal's forces. Next, Scipio took the fight to the enemy by threatening Carthage and drawing Hannibal into a fight at Zama (202 BC) where he was decisively defeated. Scipio's scouts would have helped him to choose the ground that placed his forces at best advantage.

Julius Caesar and the Gallic Wars

Having formed the First Triumvirate with Pompey and Crassus and having become a consul, Gaius Julius Caesar (100–44 BC) saw an opportunity to further extend his power and influence by taking command of the provinces of Illyricum, Cisalpine Gaul, and

Transalpine Gaul. He commanded four legions and conducted campaigns against the Belgae in the north and the Vencti and Aquitani in the west and south-west and defeating the Gaullish chieftain Vercingetorix (82–46 BC). He crossed the channel to Britain partly to prevent aid from being sent to northern Gaul.

One of the major characteristics of Julius Caesar's campaign in Gaul and his overall style of command was *celeritas*, or speed. Once he had the information about enemy positions or movements that he required, he acted decisively and fast. He was, therefore, able to keep the initiative, take tactical opportunities and maintain the element of surprise.

As the following examples demonstrate, Caesar's ability to assess the tactical situation and act swiftly depended to a large degree on him receiving accurate and timely information from his scouts and from other reconnaissance activities.

When he encountered the Helvetii (58 BC), Julius Caesar was up against an enemy for which he had particular respect among the

Julius Caesar receiving the surrender of the Gallic Chief Vercingetorix after the battle of Alesia (52 BC). (Lionel Boyer/Wikimedia Commons)

tribes of Gaul. The Helvetii were battle hardened due to their almost constant struggle to keep German tribes out of Gallic territory. They also had ambitions to spread from their confined territory near the Rhine and the Jura mountains, and even to conquer the whole of Gaul.

When Julius Caesar heard that the Helvetii were on the march and requested to pass through the Roman province, he ordered the construction of walls and moats before informing the Helvetii that they could not pass. The Helvetii found a way around the Roman defences and continued to threaten other tribes.

Once the Helvetii had reached the river Arar (now called Saone), it was Caesar's scouts who informed him that they were starting to cross and gave him the accurate information that three quarters of the Helvetii force had crossed, while a quarter still remained on the far side. Julius Caesar rapidly gathered three legions and marched to intercept the quarter that had not yet crossed the river. Taken by surprise, the Helvetii force was either killed or forced to flee.

With typical speed, Julius Caesar ordered a bridge to be constructed so that his army could cross the river and confront the larger Helvetian force. Having failed to negotiate a truce, the two forces moved apart, though Caesar sent his cavalry to keep an eye on Helvetian movements. On occasion, the Roman vanguard got into fights with the Helvetian rear-guard, but the pursuit continued for about a fortnight.

It was Julius Caesar's scouts who then informed him that the Helvetian force had come to a halt near some hills about eight miles away. Julius Caesar then sent a scouting party to carry out a more detailed reconnaissance of the ground and how the Roman forces might best approach. Once he had received word from the scouting party that the approach should be relatively easy, he sent two legions under Titus Labienus to scale the height and take the high ground. He then ordered a well-regarded officer, tribune Publius Considius, to advance with the scouts.

It was imperative in Julius Caesar's plan of attack that his force should only attack once Labienus was in possession of the high ground. Labienus did in fact succeed in attaining the high ground, and yet Publius Considius, seeing men on the heights, mistook them for the enemy and reported this back to Caesar, who then withdrew his own force to a safe place. Meanwhile, Labienus awaited sight of Caesar's force before joining what had been planned as a simultaneous attack. It was only when scouts sent by Caesar confirmed that it was Romans on the height, and not Helvetians, that the mistake was realised. By then it was too late, and the opportunity had been missed.

It is notable in the account that Publius Considius alone was held to be at fault and not the scouts. Considius is described in *The Gallic Wars* as reacting 'in sheer panic'. Otherwise, Julius Caesar retained faith in his scouts and the accuracy of their cool-headed reports.

During his campaign against the Belgae which began in 57 BC, Julius Caesar was again fortunate to be forewarned by his scouts of a coming attack by a very large force. Caesar moved back rapidly across the river Axona (present day Aisne), and made a camp. The only bridge across the river was protected on the far side by six cohorts and a large rampart and ditch.

Caesar then strengthened his position, placed his artillery to protect his flanks and positioned his front ranks just in front of a marsh that the enemy would have to cross in order to reach them. After several clashes with the enemy, the Belgae decided to withdraw in such a way as to make it look like they were fleeing. Caesar's scouts reported this, and it was then confirmed by further reconnaissance. In order to avoid an ambush, Caesar did not attempt to follow immediately but sent cavalry in the morning with three legions behind them to attack their rear-guard. This caused a rout among the enemy, without any danger to the Romans of being compromised.

Through the accuracy of his reconnaissance, Caesar had again been able to gauge the state and movement of the enemy's forces and order an attack that minimised the risk to Roman forces, while causing maximum damage to the enemy.

Caesar's speed of action was once more facilitated by his scouts during the siege of Cenabum (52 BC), a fortified town near the River Loire. Having pitched camp in front of the town, as it was getting late, he decided to wait until the following day to make the assault.

During the night, the men of the town of Cenabum moved out and this was reported to Caesar by his scouts. This gave Caesar the opportunity to invest the town with the legions that had already been placed on standby.

It was not just the Romans who during this campaign used scouts to keep themselves apprised of enemy movements. Near Avaricum, an oppidum near what is now the city of Bourges in central France, while following Caesar, the Gallic chief Vercingetorix organised scouts to cover the approach of the Roman army in teams, covering different parts of the day. Knowing that the Roman foraging parties had to roam far and wide to get enough corn, he used his scouts to observe them and inflicted heavy losses when the Romans were most exposed.

Cavalry Scouts

In his work *Hipparchicus,* Xenophon recommends that cavalry scouts should be sent out in all circumstances in advance of an army, whether in friendly or hostile territory, to identify the best routes and keep the army from getting held up in difficult terrain or areas where they might be in danger of ambush. Their other obvious duty was to identify enemy positions so that the general could make tactical choices and prepare and position his forces in advance of contact.

The cavalry units that might be used for such work in the Roman army were the *cohorts equitate,* which were a combination of cavalry and foot soldiers whose deployment depended on the terrain. Apart from topographical and tactical scouting, the *cohorts equitate* would also provide some measure of protection for the advancing main force, positioned as they were not only ahead of the force but also often on the flanks and behind, as Vegetius recommends:

> The general, before he puts his troops in motion, should send out detachments of trusty and experienced soldiers well mounted, to reconnoitre the places through which he is to march, in front, in rear, and on the right and left, lest he should fall into ambuscades.

The *cohorts equitate* would also identify suitable places for the army to camp, taking into account the tactical position, as well as access to essential water supplies and potential for forage.

Failure to provide a scouting screen could put an army in extreme danger of surprise and ambush. Although the reasons for the Roman predicament in the battle of the Teutoberg Forest (AD 9) are complex, it is notable that the Roman commander Publius Qunctilius Varus (49 BC–AD 9) failed to send out any reconnaissance parties that might have provided advance warning of the presence of the Goths, and given the main force more time to prepare.

However, when Germanicus invaded Germania in the autumn of AD 14 in order to settle the score, he took care to place his cavalry and auxiliary cohorts ahead of the main force to provide ample warning of any impending attack.

Good intelligence is one thing; acting on it is another. Caesar shows in *The Gallic Wars* the textbook way of combining good intelligence with rapid decision making and action. Prior to the disaster at the battle of Adrianople (AD 378), however, although the Roman Emperor Flavius Julius Valens Augustus had been alerted to the presence of a Goth army in good time, his failure to take appropriate action contributed to perhaps the greatest defeat for Roman arms.

CHAPTER 2

Medieval Warfare

Some historians of the medieval period have not given much space to the work of reconnaissance in battle tactics or, as in the case of Sir Charles Oman, have positively dismissed it as part of the tactical toolkit of the medieval commander. The following are some examples of battles where the historical record suggests reconnaissance and scouting played a significant role.

Reconnaissance in Battles and Campaigns of the Medieval Era

At the battle of the Catalaunian Fields, Attila the Hun (AD 453) clashed with a Roman army under General Flavius Aetius (AD 391–454) allied to the Visigoths under King Theodoric I. The battle was a significant victory for the Romans, forcing the Goths out of Gaul. The centrepiece of the battle was a ridge, which the Roman commander occupied in order to observe the deployment of the Huns before offering battle. Attila, on the other hand, was most probably advised by his own scouts of the strategic importance of the ridge and he therefore gave orders that his forces should attempt to take it.

At the strategically important battle of Vouillé in the spring of 507, which marked the extension of Frankish power throughout France, the Frankish leader Clovis (AD 466–511), who had already driven the Romans out of Gaul, prepared a base near Soissons before taking on the Western Goths (Visigoths) under Alaric II (AD 484–507). His ambition was to invade Aquitania and conquer South-West Gaul. After preliminary diplomatic manoeuvres and alliances with the Visigoths, Clovis finally showed his hand by crossing the river Vienne and making camp near Vouillé, north-west of Poitiers. At this point, he sent his scouts out to pinpoint the location of the Visigoth army. This enabled him to select an optimum site for battle and to arrange his army accordingly. Although details of the battle itself are vague, whatever disposition Clovis chose proved to be successful and the Frankish phalanx resisted Visigoth attempts to break it. Clovis then moved on to occupy Bordeaux and the future of Gaul was sealed. Thereon it would be a country dominated by the Franks.

Battle of Roncesvalles Pass

In 778 the Emperor Charlemagne, leader of the Frankish forces, was returning from an unsuccessful foray to capture Zaragoza. He had sacked Pamplona, and his forces had ravaged several towns in the Basque region. Charlemagne then led his force through the pass of Roncesvalles, in the Pyrenees mountains.

Unknown to the Frankish forces, they were being shadowed all the while by Basque scouts. The Basques set up a massive ambush, which cut off the rear of the Frankish force, attacking the baggage train, its protecting force and the rear-guard, which included a knight called Roland whose name would be immortalised. The battle was an example of how effective guerrilla-style warfare could be, especially against an army trained to fight in open battle. However, the Frankish commanders did not seem to learn the lesson offered by the ambush, and another Frankish force suffered a similar fate.

Defence of Constantinople

After repeated attempts to conquer Constantinople, the task was taken on by the Caliph Sulayman ibn Abd al-Malik in 715. The

Muslims were confident of victory, as a prophecy had foretold that a Caliph bearing the name of a prophet, in this case Solomon, would finally take the city.

The vanguard of Sulayman's army, under the command of Maslama ibn Abd al-Malik, moved into central Anatolia in the spring of 716, while the main force remained in Syria. The Muslim fleet meanwhile waited off the coast of Pamphylia. Constantinople, under the newly installed Emperor Leo III the Isaurian, awaited its fate. When the Muslim attack was finally unleashed, it failed, the Muslim fleet literally dashed by a storm against the city walls and then finished off by the Byzantine fleet.

After this, the Muslims changed their tactics, focusing on looting raids and avoiding major battles. Similarly, the Byzantines adopted a skirmishing strategy, keeping a watch on Muslim forces and shadowing them. Here, there was ample opportunity for reconnaissance forces to keep an eye on enemy movements and report back to local commanders, who would then choose their moment to concentrate forces and intercept and attack the marauding Arabs.

Battle of Hastings

In 1066, Duke William of Normandy (1028–1087), later known as William the Conqueror, invaded England to claim what he regarded as his right to the English throne. He also had the support of the Pope Alexander II who regarded the invasion as a crusade against the schismatic English. King Harold Godwinson II (*c.* 1022–1066) had his work cut out defending his kingdom almost simultaneously from an attack in the north from the Norwegian King Harald Sigurdsson 'Hadrada' (1046–1066) along with Harold's younger brother Tostig as well as from Duke William in the south. The battle of Hastings is an interesting example of the use of reconnaissance and scouts, not least because the scouts on both sides are depicted within the famous Bayeux Tapestry and, unusually, one of them is named. In the tapestry depiction, the scout named Vital returns to inform Duke William of the sighting of the English army under Harold.

Vital informs William of Normandy of the presence of the English army near Hastings. (Bayeux Tapestry)

This vital intelligence gave William ample time in which to prepare his army and organise them for the coming clash.

William's awareness of the approach of Harold's army appears to contrast with the recent battle of Stamford Bridge, near York, on 25 September, where the Norwegian King Harald Sigurdsson did not seem to have been aware of the approaching English army, was caught by surprise and defeated.

Duke William received his warning on the evening before the battle, 13 October, giving him even more time to prepare. It also appears that at the time of the scout Vital's return, William's army was out foraging. For an army to be caught by surprise while many of its units were foraging would put it at an obvious disadvantage. William, therefore, was given plenty of warning to call back his troops from their foraging expedition and prepare for battle. This is all the more significant given the pace of Harold's approach. Harold had returned with great speed from the north to meet the new emergency, and he would have not been inclined to give the Normans any time in which to establish themselves.

The high value that Duke William placed on reconnaissance is attested by his personal scouting foray with a retinue around the Hastings area, first on horseback and then on foot, due to the difficulties of negotiating a path. During the recce, he is said to have carried the hauberk of one of his knights, who was perhaps unused to walking or who may have been exhausted by the extent of the reconnaissance.

As the Norman army advanced the following day, 14 October, the Norman scouts located at Hedgland on Telham Hill, spotted the English army two miles away, positioned on what has since been called Battle Hill. The Normans then engaged with the English army at about 9 a.m. It is significant that, once the English army had been spotted by the scouts that morning, the rest of the Norman army moved with decisive speed to engage the English. They were able to do so because they already had accurate information on the presence and likely size and formation of the English army provided by their reconnaissance, and William had already worked out his tactics for the battle and given his orders. The speed of the Norman

An English lookout informs Harold of the advance of the Norman army. (Bayeux Tapestry)

advance, moreover, had the effect of putting the English on the back foot. The English position on Battle Hill was not chosen at leisure, but on the spur of the moment as Harold saw the Norman army advancing against him. He was fortunate in the circumstances to be able to choose a tactically sound defensive position.

Although Harold's plan would undoubtedly have been to come upon William before he was fully established on English soil, and in a country with which he was unfamiliar and unwelcome, the opposite proved to be the case. As Harold's forces began to move from their camp in the morning, it was the Normans who were moving like an oiled machine against them, forcing the English to make hasty decisions. Reports of the battle even suggest that Harold did not even have the time to assemble his full force on Battle Hill before the Normans were upon them.

Whatever may be said about the decisions and tactics during the course of the battle itself, it appears that, through good reconnaissance and preparation, William of Normandy had the initiative from the start.

El Cid and the Reconquista (711–1492)

Having been the champion of Sancho II of León and Castile, Rodrigo Díaz de Vivar (1043–1099), commonly known as El Cid, soon fell out of favour with his successor, Alfonso VI, and was banished. He went into service with the emir of Zaragoza, al-Mu'tamin. During his exile El Cid (the name being derived from the Arabic *al-sayyid* or 'Lord'), continued to build a reputation as a warrior.

While working as what may be called a mercenary, El Cid led his forces through enemy occupied territory from the summer of 1081 with no secure base to fall back on. He maintained his force through his competence as a commander and ability to find enough forage and booty to keep the troops interested.

After the capture of Toledo in 1085, the gradually expanding Christian Spain was faced with a new threat in the form of Yusuf ibn Tashfin, leader of the Berber Moroccan Almoravid empire, who crossed the Straits of Gibraltar, landing at Algeciras with

4,000 Berber warriors. Yusuf ibn Tashfin moved to Seville, where he declared a jihad, or holy war, and then north to Badajoz near the border with Portugal. Alfonso met Yusuf ibn Tashfin at the battle of Sagrajas near Badajoz, in October 1086 and was defeated. Humbled, the king turned to his former champion, the one Christian commander in Spain who remained undefeated in battle.

Meanwhile, El Cid had set his sights on Valencia, which he eventually captured in 1094.

Although it is difficult to discern the details of El Cid's use of scouts and reconnaissance from the available sources, his respect for, and use of, advanced reconnaissance techniques is implicit in the manner with which he conducted his campaigns. El Cid became known in Spain as *El Campeador*, which could mean military teacher or expert. El Cid's success was not due just to his heroic bravery or good luck, but to careful planning, study of the ground, analysis of the weak points of the enemy and of any other potentially favourable conditions, including the weather. El Cid's ability to marry both theory and practice along with prudence and decisive action helps to explain his success.

The battle of Cuarte outside Valencia in October 1094 is a good example of El Cid's careful consideration of a variety of conditions in preparation for the engagement. Knowing that the Almoravids were approaching Valencia to lay siege, El Cid arranged for Moorish allies who were opposed to the Almoravids to lead the Almoravid quartermasters to an irrigated meadow nearby at Cuarte, which featured orchards and a network of irrigation channels and ditches. El Cid knew that heavy autumn rains were a common occurrence and he arranged for the irrigation system and channels to be breached at the appropriate time. As he readied his troops inside the city, El Cid was advised by observers that birds had been seen flying low to the ground, one of the signs of approaching rain. It was also Ramadan and El Cid timed his foray from the city to coincide with the time when Muslim sentries would be least alert. Dividing his forces, El Cid led them out of the city ready for the assault, one part of his force heading to the Almoravid camp, while another readied to attack the main force outside the city. The double attack on the Almoravids, timed with torrential rain and the breaching of the canals, caused immediate confusion. The Almoravid camp was thrown into chaos, while the Almoravid force near the city believed that they were being attacked from behind and that all was lost.

El Cid's careful evaluation of the many contributing factors before battle and his decisive action at the opportune moment, puts one in mind of Julius Caesar, and there is little doubt that they had equal respect for the work of scouts and reconnaissance forces.

As the conflicts between Christians and Muslims continued, the Christians adopted some of the Muslim practices, including their use of light highly mobile cavalry, in addition to the more familiar Christian heavy knight-carrying destriers. The light cavalry were known as *adalides,* and they soon became masters of frontier warfare. They had an intimate knowledge of the ground and of enemy movements, and they could both bring news to the commander of an enemy advance, or position themselves in the most effective places for an ambush. During the reign of Alfonso XI (1311–1350), some of his most trusted knights served in this light cavalry, a testament to the importance of this mobile reconnaissance force in the eyes of the king.

Battle of Tagliacozzo (1268)

This battle near Tagliacozzo in southern Italy was the last act in the attempt by the Hohenstauffen line to extend their power in mainland Italy. Their defeat at the hands of Charles of Anjou led instead to the extension of Angevin power in southern Italy. When the forces under Charles of Anjou and Conradin of Hohenstaufen met, they both had to negotiate hilly countryside and take into account the tactical features, notably the river Salto. Any medieval commander knew that when crossing a river an army was at its most vulnerable, and, on this occasion Charles of Anjou's forces included between 3,000 and 5,000 cavalry, including French and Provencal knights. The opposing force under Henry of Castile and Conradin consisted of between 5,000 and 6,000 German and Castilian cavalry along with Italian mercenary knights.

For cavalry forces of this size to manoeuvre effectively and take up optimum tactical positions in such varied and difficult terrain required a constant flow of information from both spies and scouts, reporting back to commanders not only on the lie of the land and any potential ambush sites, but also the movements of the enemy forces. Such intelligence was even more vital in view of the fact that cavalry forces could cover ground at great speed, leaving little time to make decisions.

Battle of Tannenberg/First Battle of Grunwald (1410)

Huge multi-ethnic forces were involved in this battle, including both the Polish and Lithuanian armies rallied against their common enemy, the military monastic order of the Teutonic Knights, bringing to an end their crusading ambitions to expand eastwards and marking the rise of Poland and Lithuania in eastern Europe.

High levels of communication and planning would be required to coordinate the movements of the allied armies before they moved against their formidable foe. Scouts would have been working in front of both armies to get the best idea of the lie of the land and enemy movements. On 15 July, the Polish-Lithuanian forces were informed by the scouts that the Teutonic main force was only two miles away, near the villages of Grunwald and Tannenberg. The conditions were misty, so accurate intelligence was vital. The ground was boggy, making it tricky for cavalry, and the Teutonic knights had also prepared defences. Attacks by both Lithuanian and Polish forces eventually led to the defeat of the Teutonic Knights.

Byzantine and Muslim Scouts

Apart from reporting on mobile incursions, a more localised system of Byzantine scouts and sentries also kept watch on the grey frontiers between Byzantine and Muslim areas of influence. Called *vigilators* or *caminus*, these lookouts were responsible for identifying a potential threat from Muslim forces approaching the border. The border sentries would patrol the area to which they were assigned, keeping an eye on remote trails and pathways by which an enemy might attempt to approach, as well as areas that might be suitable for an enemy force to pitch camp, such as protected ground with access to water. Once enemy movement that posed a significant potential threat was identified, a relay system was set in motion, whereby a runner or rider would alert the nearest lookout station, which may have been three or four miles away, after which a rider would be despatched to warn the provincial commander.

At best speed, a horseman, or camel rider, might cover 50 miles in a day. Once the provincial commander had received the news, he would then need to make his tactical decisions about what forces to deploy, and a rider would be sent back to the frontier area to advise the local forces of the plan.

An alternative means of alerting rear forces of an invasion would be a system of beacons lit on prominent hills. When one beacon was lit on the frontier, observers at the next beacon station would then light another and so on until a beacon was seen from Constantinople itself.

When Byzantine forces were advancing into enemy territory, they commonly sent experienced scouts known as *skoulkatoures* and *minsouratores*, who would use their expertise to work out the best routes, avoiding potentially compromising areas and identifying sites where the advancing forces could safely camp. These scouts would be lightly armed with swords and wore no armour.

The Byzantines also deployed mounted scouts known as *prokoursatores* who were usually armed, providing the ability to defend themselves if necessary, bearing in mind that their main task was to gather information and get back swiftly to warn the force commander, so that he could make effective tactical decisions to respond to the unfolding developments.

Accounts of the time, such as the *Tactica* of the Byzantine Emperor Leo VI, frequently stress the need for experienced men to act as scouts. Ideally, they would be local men with the natural instincts of countrymen and hunters for good cover and fieldcraft along with weather conditions. Not only was knowledge of the ground vital, they also needed the tactical skill to make accurate assessments of enemy numbers. This was not as straightforward as it might at first seem because enemy commanders might arrange their forces in such a way as to disguise their numbers. A force stretched out over a wide area with gaps between units might have seemed larger than it was. Conversely, a large force bunched together might have seemed smaller than it was. Based on this information, the local commander would make his decision to either stand and fight, withdraw and await reinforcements, or wear down the enemy with harrying tactics. If the decision was to stand and fight, the commander would also be dependent on his scouts to choose the right ground.

The small, fast-moving cavalry forces that shadowed Muslim forces threatening Byzantium during the 10th and 11th centuries were in themselves a form of scouting unit. It was important not to lose touch with the enemy but also to keep enough distance so that the lightly armed shadowing forces were not themselves annihilated. However, the shadow force also had enough power available to attack if the opportunity presented itself. This guerrilla style of warfare, relying on good intelligence, guile and communications, was a sea change from previous methods of warfare which relied on large forces in open battle.

What was good for the Byzantines was also appreciated by the Muslims. The *Tafrij al-Kurun fi Tadbir al-Hurub* (*Manual of War*), written by the Muslim scholar Umar Ibn Ibrahim al-Awsi Al-Ansari, describes a cavalry scouting party called the *al-tali'ah,* who would precede the main force and provide advance information on enemy movements. The same requirements for skilled, courageous and experienced men to form such scouting groups is also underlined. Their initial advice, based on their previous military experience, would enable the commander to plan his movements and organise his forces.

CHAPTER 3

The Revolutionary Years

While European powers jostled for position on the European continent during the 18th century, including the War of the Spanish Succession (1701–1714), the War of the Austrian Succession (1740–1748) and the Seven Years' War (1756–1763), France and Britain also became increasingly competitive and ambitious in their colonial expansion in the American colonies. Although the British attitude had initially been to let the colonists battle it out among themselves, increasing French aggression in response to British expansion westwards made it clear that the colonists would require both military and naval support.

During the campaigns of what was called the French and Indian War (1754–1763), in North America, which started in the Ohio valley, both sides had native Indian allies from whom they learned new ways of conducting warfare. It was not the native Indian style to stand in the open in brightly coloured uniforms in order to become easy targets. Never having encountered an enemy such as the Indians and having been trained to march in columns on European battlefields, the British realised that they would have to adapt to the new environment if they were not to lose the war.

Gorham's Rangers (1744–1762)

This unit was formed largely to cope with resistance by Arcadian and Mi'kmaq colonists and natives in Nova Scotia. Realising that their best hope was to fight their opponents using their own methods, the British formed a company of native Americans in Massachusetts led by John Gorham. Due to its success in reconnaissance and small-unit operations, the unit was expanded with royal approval. As time went on, the unit recruited Anglo-Americans and a uniform was developed which may have been either blue or grey. This was a recognition that scarlet was not appropriate for covert work. Reconnaissance operations mounted by Gorham's Rangers included a recce of Louisbourg harbour in 1745, when the rangers disguised themselves as fishermen and approached the harbour in a fishing boat. Having discovered that powerful French reinforcements were due to arrive, they reported back to their headquarters and averted what might have been a costly attack on the harbour.

Rogers' Rangers

Robert Rogers (1731–1795) was raised in Massachusetts, his family having bought land at least two miles from the nearest settlement. Their farming homestead was in an area populated by Indians, some of whom were hostile to white colonists. Close proximity to the Indians helped Robert to learn their ways of hunting, tracking and warfare. As the tension between French and British settlers grew, the Rogers family was forced to abandon their farm and move to the main settlement at Methuen where Robert joined the local militia, aged only 17. He served in Captain David Ladd's scouting company in 1747 and in Ebenezer Eastman's scouting company. Rogers then moved to Portsmouth, New Hampshire, where he began to recruit a unit of his own. He then led his unit of rangers on forays against French settlements, often appearing by surprise, having travelled long distances on foot, crossing frozen lakes and using snowshoes in deep snow. The success of Rogers' operations led to the demand for more such forces and the corps grew to over 1,000 men.

The French and their Indian allies ran similar operations, and on 21 January 1757, Rogers and a group of his rangers were

Robert Rogers. His Ranger company would have a lasting influence. (US Army Center of Military History/Wikimedia Commons)

ambushed by French forces, militiamen and Indians under French ensign Joseph de la Durantaye near Lake George. A battle raged inconclusively until sunset, upon which Rogers ordered his men to slip away under cover of darkness. The French reported that the British unit had an advantage over them as they were wearing snowshoes. This is a testament to Rogers' attention to detail and the importance of the right equipment for special operations. However, several of Rogers' men were captured and sold into slavery. On this occasion, Rogers had broken one of his own rules by ordering his unit to retreat by the way they had come. This gave the opportunity to the enemy to set up an ambush.

Rogers had devised a set of rules, known as The True Plan of Discipline, for his men to follow, which were designed to provide best protection for the unit as a whole and for individual soldiers in enemy territory. The rules included advice on marching order and how that should be adapted to the different terrain and conditions. The men should 'march abreast' on soft or marshy ground before returning to single-order file on firm ground in order to reduce an enemy's ability to track them. The unit should only set up camp once it was dark and only in a place where sentries had a good view of the surrounding area. Point 4 of Rogers' rules dealt with reconnaissance:

> Sometime before you come to the place you would *reconnoitre*, make a stand, and send one or two men in whom you can confide. To look at the best ground for making your observations.

He also recommended that 'on your return take a different route from that which you went out'.

His advice on firing drills was revolutionary for his time. When coming under fire, the men were to 'fall, or squat down, till it is over; then rise and discharge at them'. They were also advised to 'advance from tree to tree'. This kind of manoeuvring would be familiar to a modern soldier but was novel for a regular soldier of the 18th century.

Rogers was also aware of the importance of the morning and evening stand-to, which is still a drill in modern infantry tactics:

> At the first dawn of day, awake your whole detachment; that being the time when the savages choose to fall upon their enemies.

There is advice on avoiding the usual fords across rivers and on keeping clear of lakes. There are also recommendations on maintaining contact between boats on a river.

All these rules amount to a coherent guide for special operations and reconnaissance which have stood the test of time. So much so, that an adapted version is still part of the creed of the US 75th Ranger Regiment.

Morgan's Riflemen (1775–1777)

Alongside the development of long-range reconnaissance units, such as the rangers, there were also developments in light infantry tactics such as sharpshooting and their weaponry. Colonel Daniel Morgan, a pioneer, soldier and politician from Virginia, formed an elite light infantry unit which was named after him. The unit was composed entirely of marksmen who had to pass a demanding shooting test. At the battle of Saratoga in 1777, which proved to be a turning point in favour of the colonists during the American War of Independence, Morgan's riflemen demonstrated their sharpshooting skills to considerable effect. Well-placed and concealed sharpshooters took their toll of senior British officers as well as Indian guides acting on behalf of the British, causing chaos and a rapid decline in British morale.

Morgan's riflemen were not just a group of elite marksmen; they were also highly mobile and expected to travel long distances with everything they needed and with the ability to fire accurately and effectively once they arrived.

The rifle they carried was originally called the Pennsylvania rifle and was later known as the Kentucky rifle. The 889 mm barrel was rifled, contributing to greater accuracy.

Accurate fire by the American colonial army often made up for their lack of numbers and the threat of well-drilled British bayonets. As ingenuity and accuracy over regulated discipline became more familiar, the seeds of modern infantry tactics were sown.

Daniel Boone (1734–1820) was a legendary frontiersman and pioneer who roamed far and wide while tracking, hunting and exploring. He was one of the first to find a way through the Cumberland Gap in the Appalachian Mountains, which enabled colonists to reach the Mississippi river. The route was called the Wilderness Road and provided access to the first settlements in Kentucky.

In July 1776, Boone led a group to rescue his daughter and two other girls who had been captured by Indians.

Ferguson's Rifle Corps

The advantages of light infantry and accurate shooting were not lost on the British. Patrick Ferguson arrived in the colonies in 1777 with a rifle corps equipped with a new breech-loading rifle – the Ferguson rifle. It had a rate of fire of between six and ten rounds per minute. The fact that the rifle could be loaded via the breach made it much easier to re-load from concealed positions. However, there were weaknesses in the design that led to frequent breakages. The Ferguson rifle inspired further developments in rifle design.

Voltigeurs

In Europe the development of light infantry also promoted a new form of soldiering, where a certain level of informality and use of initiative was encouraged. Lessons learned in theatres of war, such as the American colonies, demonstrated the advantages of uniforms that were designed to blend in with the landscape, rather than attract notice.

The name *voltigeur* given to the French skirmisher units created by Napoleon in 1804, literally meant 'jumper'. The idea was that the nimble young men would literally jump on to the back of cavalry horses in order to increase their mobility, though it soon became apparent that this was not very practical. The *voltigeurs* received special training in marksmanship and also in the art of using cover and taking the initiative. Operating in a loose formation, their skirmish line had the effect of screening the main battalion from the enemy. So successful were the *voltigeurs* that their numbers were increased in 1807 and they were honoured with a position on the left of the line as one of the flank companies.

Chasseurs à cheval and *Chasseurs à pied*

The *chasseurs à cheval* were light cavalry, distinguished by a green uniform that was less showy than that of the other cavalry regiments.

The *chasseurs à pied*, like the *voltigeurs*, were taught marksmanship and also wore darker colours than the infantry of the line. They were roughly equivalent to British light infantry and German *Jäger*. Like their British and French counterparts, there was strong emphasis on fieldcraft and individual initiative. Orders were given by bugle, rather than by drum, and they usually attracted the best officers. They often operated alongside the *voltigeurs* to form scouting parties to gather essential information about enemy dispositions and movements.

Jäger

Frederick the Great of Prussia employed guides and scouts from 1740. In 1744, a *Jäger* infantry branch was formed. Their tasks included scouting duties, reconnaissance and sharpshooting. *Jäger* were often well acquainted with the use of rifles. Like their counterparts in the French and British armies, *Jäger* were often used for screening and skirmishing duties. As was to become characteristic of selection for these units, soldiers with the most energy and initiative were selected, as the roles often involved making quick decisions on the ground away from the direct lines of command.

The effectiveness of the *Jäger* sharpshooters was demonstrated at the battle of Wavre, on 18–19 June 1815, when 17,000 Prussian light infantry held down 30,000 French troops under Marshal Grouchy, allowing time for the main Prussian force to go to the aid of Wellington at Waterloo.

British Light Infantry

It is no surprise that one of the main proponents of light infantry in the British Army, General Sir John Moore, had served in North America in the 82nd Regiment of Foot between 1778 and 1781. After service in locations as varied as the West Indies, Ireland and Egypt, in 1803 Moore was put in command of the Shorncliffe Army Camp, near Folkestone, where he was instrumental in training the first effective regiments of light infantry.

An 'Experimental Corps of Riflemen' had been raised by Colonel Coote Manningham and Lieutenant-General Sir William Stewart, GCB, in 1800. Dressed in dark green with black linings in order to make them less conspicuous to the enemy, they were issued with Baker rifles. The design of the Baker was influenced by the German *Jäger* rifle and by the requirements of the British rifle regiments, who needed a rifle with greater accuracy and range than the standard Land Pattern Musket. The Baker Rifle would be the standard-issue rifle for British armed forces during the Napoleonic Wars (1800–1815). They were also issued with a 21-inch sword bayonet.

The model for the new light infantry regiments was the 60th Royal American Regiment, later called the 60th Regiment of Foot, and the German *Jäger* forces serving with the British. Chastened by their experiences against the highly effective French *Chasseurs*, the British recognised that more training was required and the experiences in North America helped to bring them up to date.

Sir John Moore used his own regiment of light infantry, the 52nd (Oxfordshire) Regiment of Foot, as the model for the new training regime at Shorncliffe. Lieutenant-Colonel Kenneth MacKenzie was put in charge of training and soon the 43rd Foot was added to the 52nd as a light infantry regiment.

Apart from the details of light infantry training, perhaps the most remarkable aspect of the Shorncliffe system was its ethos. Instead of the extreme obedience to detailed orders and lines of command prevalent in the line infantry regiments, officers and men drilled together, and the goal was mutual respect and shared values. As light infantry were expected to work in extended formations well away from direct lines of command, they had to use their initiative.

The Experimental Corps of Riflemen became the 95th (Rifle) Regiment of Foot, their duties including sharpshooting, skirmishing and scouting. In 1816 they were renamed as the Rifle Brigade.

The 95th was the model of the new kind of infantryman and foreshadowed the revolution in military tactics where accurate fire

The British light infantry wore green uniforms and used cover in a manner more associated with a modern soldier. (Wikimedia Commons)

from cover became more important. Wearing practical dark green jackets and pantaloons with black facings and cuffs, they looked as effective as they were.

The 95th would make their mark in the Peninsular War (1808–1814) where they fired the opening shots of the war after landing in Mondego Bay, near Figueira da Foz. Later, the 1st Battalion of the 95th took part in Sir John Moore's fighting retreat to Corunna.

As the reputation of the 95th spread and more volunteers came forward to make up for losses at Corunna and other battles, a third battalion was formed. All three battalions of the 95th took part in the key tactical battles of the Peninsular War, including the battle of Buçaco in September 1810 and the withdrawal to the prepared Lines of Torres Vedras. They were also employed as line infantry during the costly storming of the fortifications at Badajoz.

Retreat to Corunna (December 1808–January 1809)

After a failed attempt to defeat Marshal Soult's force near Madrid on 21 December 1808, and with a growing threat from Napoleon's army moving up from the south of Spain, Sir John Moore decided on a fighting retreat to the port of Corunna, in north-west Spain.

The retreat began in December 1808 in severe weather. With the enemy snapping at his heels, Moore deployed the Light Division as rear guard to keep the enemy at bay. Set-piece battles were continuously fought between the Light Division, British light cavalry and the French advance guard. On one occasion, a French general was shot at long range by a member of the Light Division with a Baker rifle.

Retreat to Lisbon

The reconnaissance skills of the Light Division were demonstrated in 1810, as the British awaited the arrival of the French army under Marshal André Masséna. Based on Almeida, Portugal's defensive bastion in Riba-Côa, in the north-western hills of Portugal near the border with Spain, the Light Division, commanded by Major-General 'Black Bob' Craufurd, set up a series of outposts to provide early warning of French movements and a screen against enemy attempts to gain information about British positions. The tactics proved highly effective. While the French were unable to penetrate the British screen to gain information, the Light Division passed back accurate information about French movements and dispositions to senior commanders. When the French sent patrols in greater force, the Light Division soldiers in the area deployed rapidly to provide a riposte. The light division along with Portuguese elite light infantry *Caçadores*, Royal Horse Artillery, Dragoons and Hussars numbered about 5,000. Their success, however, may have gone to Craufurd's head, for when the French attacked in force on 24 July 1810, rather than fall back over the Côa River, as he had been ordered to do, he chose to take them on. The British received a severe drubbing by the French in a precarious position, with a river at their backs, and only the fine fighting qualities of the men of the elite Light Division enabled them to escape a disaster. Craufurd was lucky to receive a comparatively mild rebuke for his misjudgement from the Commander-in-Chief, Arthur Wellesley.

The Lines of Torres Vedras (October 1810)

As Marshal Masséna entered Portugal with an army of over 60,000 men, under orders from Napoleon to drive out the British army and take Lisbon, unknown to them, the Duke of Wellington had already made plans well in advance for their reception. A series of redoubts and other defences were being built in the municipality of Torres Vedras, about 40 miles northwest of Lisbon. Built under the direction of Lieutenant-Colonel Sir Richard Fletcher, 1st Baronet, of the Royal Engineers, the series of redoubts, batteries and ramparts were arranged in three lines between the river Tagus and the sea, taking advantage of the natural defensive properties of the hills. The first line ran through the town of Torres Vedras itself. The second was about six miles behind while a third was based on the fort of St Julian on the coast between Lisbon and Estoril and was designed to protect an emergency embarkation of troops. Wellington's plan was to fall back on these hills using scorched-earth tactics to delay the French. He also planned to fight a delaying action, and, for this, he chose a ridge in the hills of Buçaco north-east of Figueira da Foz, where the British had first landed in Portugal. This would block the French army as they headed south towards the university city of Coimbra.

Wellington deployed his troops on the reverse, western side of the ridge near the convent of Buçaco and ordered his engineers to build a communications road along the ridge so that units could be rapidly deployed wherever they were needed.

Confident in their superior numbers and with the impression that the Anglo-Portuguese army was on the run, the French commanders Marshal Masséna, Marshal Michel Ney and General Jean Louis Ebénézer Reynier could not resist the temptation to defeat the allies once and for all, though Reynier later changed his mind. Initial French reconnaissance offered no information to dissuade them. Wellington ordered the Anglo-Portuguese army to keep a low profile. They were not allowed campfires during the night and, therefore, French recce patrols would have seen little evidence of a substantial force.

When the French attacked up the hill on 27 September, they were met with withering fire from British and Portuguese units that rose up from the ground as the French approached the crest. Volleys of fire were followed by bayonet charges that sent the French running back down the hill. General Craufurd, with the 52nd Light Infantry, positioned at the north end of the hill, lay low behind a rock and then stood up as the French appeared calling out to his men to avenge the death at Corunna of their previous commander, General Sir John Moore.

As the French probed to the north to attempt to outflank the British position, Wellington ordered the army to retreat, with the Light Infantry covering the rear. Disoriented by bad maps and suffering from lack of food, the French continued southwards to eventually find themselves in front of a line of hills from which their enemies could fire at them with muskets, rifles and cannon from seemingly impenetrable redoubts in front of which all available cover had been carefully cleared. There was a sophisticated communications system between the fortifications to warn of an enemy advance. The French attempted to push through the lines at Sobral on 13–14 October but were repulsed. As his army became increasingly frustrated and hungry, Marshal Masséna decided to retreat north on 14 November to find food and forage. In March

1811, he began the full-scale retreat to Spain, leaving the way clear for the Anglo-Portuguese army to advance to Badajoz and Ciudad Rodrigo.

Deployment of Scouting Parties and Reconnaissance by British Light Infantry

When on the march, a battalion or company would put out an advance guard which would be in regular touch with the main body of troops as they advanced.

The light company would advance 500 paces in front of the main body in clear weather. They would be closer in foggy weather or at night. At 200 paces beyond them would be a section, and 100 paces beyond that, a sergeant and six men. Two other sections would be positioned to the sides, from which a sergeant and six men would advance 100 paces in the same manner as the central section. From each of the three most advanced sections of seven men, individual scouts would probe forward.

In this way, an ordinary light company was sensitive to any enemy movements at enough distance to give them time to send warnings back to command and for commanders to make any necessary tactical decisions. When necessary, the extended reconnaissance arms could be withdrawn, for example, when a major enemy force was encountered.

The Campaign of France (January–March 1814)

After the battle of Leipzig, 16–19 October 1813, French home borders were under threat from three coalition armies under Karl Philipp, Prince of Schwarzenberg, Gebhard Leberecht von Blücher, *Fürst* (sovereign prince) von Wahlstatt, Charles John, Crown Prince of Norway, and Count Levin August Gottlieb Theophil Graf von Bennigsen of Germany.

Not only was Leipzig a personal disaster for Napoleon as his first major defeat, it also severely depleted his forces, aggravating the losses from the disastrous retreat from Moscow in the winter of 1812. While he urgently recruited a new army of young and inexperienced conscripts, Napoleon also recognised a need for light cavalry to carry out reconnaissance and keep an eye on the allied forces approaching the French border.

The French had long been aware of the effectiveness of the Cossacks with their ability to scout, infiltrate and carry out hit-and-run operations. From as early as 1805, Napoleon had planned to create a similar force to counter the Cossacks and to carry out scouting and raids. The new unit, broadly named as the Éclaireurs of the Guard, were to be light, swift, able to carry out raids and to perform reconnaissance. The other inspiration for these new units was 1st Polish Light Cavalry Regiment of the Imperial Guard. Some of the scouts would be equipped with Polish style lances. This was not the first time that Napoleon had thought of creating such a unit. In 1806, Napoleon had suggested the creation of a unit of mounted scouts, or *Éclaireurs à cheval*, mounted on the hardy horses of the Camargue that could stand all weathers.

Lipka Tatars were also employed as scouts by the French, attached to the 3rd Regiment of the Éclaireurs of the Guard. A Polish Cossack corps was also formed.

In true scout-like fashion, the scout horses had simplified saddles and harnesses. The men's uniforms were comparatively plain by cavalry standards, with dark colours and black facings along with black shakos. The officers of the scout regiments were carefully selected for their bravery and experience. The scouts were expected to be bold and use their initiative in difficult circumstances, often under the noses of the enemy.

Formally known as the *2e* and *3e Régiments d'Éclaireurs de la Garde impériale* and informally as *Éclaireurs-Dragons* and *Éclaireurs-Lancers,* these units would play a key role in Napoleon's fast-moving tactics of the Campaign of France.

As the three Coalition armies prepared to cross the French border in December 1813, Napoleon only had about 60,000 troops at his disposal to defend the Rhine frontier, and another 30,000 in reserve to be moved where necessary. Another 60,000 under Marshal General Jean-de-Dieu Soult, 1st Duke of Dalmatia, were being gradually pushed back into northern Spain by Wellington.

Like Julius Caesar and Alexander the Great before him, Napoleon had a talent for decisive action. He needed accurate information about enemy movements in order to intercept the separate Coalition armies to best effect. Despite being heavily outnumbered and fighting three separate armies, Napoleon's lightning strikes based on accurate information about the enemy's movements caught the enemy off their guard.

The newly created Scouts of the Imperial Guard played a key role in providing Napoleon with intelligence about Coalition movements and often placed themselves in great danger. After initial intelligence revealed that the Coalition forces were coming from the east as opposed to the north, as Napoleon had expected, in January 1814 Napoleon moved with his scout units to Chalons. From here, he carried out his tactics of intrepid marches against Coalition forces, whenever and wherever they were seen in order to prevent them from combining forces.

Significantly, on 28 January, the French Scouts of the Imperial Guard had their first encounter with their opposite numbers in the Coalition forces, the Cossacks. The Cossacks were probing ahead of the vanguard of their army in order to detect French movements. The Éclaireurs of the Guard charged, forcing the Cossacks to retreat.

The next day, Napoleon's army clashed with Blücher's army at Brienne. French reconnaissance revealed that Blücher's forces were dispersed, and Napoleon took rapid advantage.

Although Napoleon defeated Blücher at Brienne, Blücher managed to get away with a substantial force, and Napoleon pushed back to La Rothière, where the French were massively outnumbered and made a retreat at nightfall, having lost 6,000 men. Undeterred,

Napoleon then moved to confront the Russians at Montreuil. Here, the scouts were employed as regular cavalry against Prussian squares. The Russians were lucky to receive reinforcements just in time. Soon, Napoleon was after the Prussians again, catching up with them at Château-Thierry. As the Prussians tried to retreat, the French inflicted 6,000 casualties on them, losing only 600 of their own.

Napoleon gave his opponents an object lesson in what could be done with a smaller, highly mobile and adaptable force against larger and more cumbersome armies. His fast interceptions of the Coalition forces were aided by using lesser-known routes, and sometimes through terrain such as marshes, which were considered impassable. Good reconnaissance and the use of local knowledge gave him the edge.

Fast-moving and at the forefront, on 5 March the 3rd Scouts rushed to secure the bridge at Berry-au-Bac, which was held by 2,000 Cossacks. After a valiant charge, the enemy were cleared from the bridge, while Napoleon looked on admiringly. After this the French continued to pursue the far larger Prussian army, though this time the Prussians withstood the attack.

Napoleon then sent the Éclaireurs of the Guard via highways and byways to the city of Rheims, where they conducted a successful covert investment of the city. From Rheims, Napoleon moved quickly to intercept Coalition communications near the river Aisne. Once again, the Éclaireurs were at the forefront, identifying the Austrian rear-guard and carrying out a swift and precise attack that caught their enemy by surprise.

After the battle of Arcis-sur-Aube (20/21 March 1814), where the French army was heavily outnumbered by the Austrian army, Napoleon realised that the writing was on the wall and that he could no longer hold back the Coalition tide that was moving ineluctably towards Paris.

The Campaign of France, and the Six Days Campaign (10–15 February 1814) in particular, burnished Napoleon's already established reputation, although, ultimately, they ended in defeat.

Whereas in previous years he had demonstrated his unparalleled skill in large-scale set-piece battles, he had now shown that he was also able to move with speed and precision with a smaller force, relying heavily on accurate reconnaissance, and destabilising and defeating his cumbersome foes before they had time to organise themselves.

Battle of Waterloo, 18 June 1815

During the 100 Days Campaign that culminated in the battle of Waterloo, and which also included the battles of Ligny (16 June), Quatre-Bras (16 June) and Wavre (18/19 June), Napoleon appeared initially to have lost none of his spark. His movement towards Brussels was rapid and caught his two foes in the area, Wellington and Blücher, off their guard. Napoleon placed himself in a position where he could confront each of his enemies separately without giving them the opportunity to combine forces. Reconnaissance information from both cavalry and spies allowed him to attempt to defeat Blücher before Wellington's Anglo-Dutch army had had a chance to establish itself.

Displaying the mastery of strategy that had made him a legend, Napoleon succeeded in concentrating his considerable forces at the pivotal point, which in this case was Charleroi. Napoleon also took care to disguise his movement, creating a diversion towards Wellington's lines of communication. By positioning himself at Charleroi, Napoleon was not only within spitting distance of both Wellington's and Blücher's armies, but also blocked the route by which the Coalition armies could come together to oppose him.

That the French army of around 125,000 men was sitting undetected right under the noses of the Coalition, suggests that Coalition reconnaissance and intelligence systems were not functioning well at this juncture.

However, from the time that Napoleon moved to defeat Blücher at Ligny, either he lost his edge, or his luck began to run out.

Napoleon carried out a personal reconnaissance of the Prussian army at Ligny on the morning of 16 June, where his opposite numbers, Blücher and Wellington, did the same. Although the French defeated the Prussians, the bulk of the Prussian army was able to escape the battlefield. Rather than maintain contact with the enemy, however, and verify their line of march, Napoleon chose to work on the assumption that they would head east towards their lines of communication. In fact, the Prussians headed north towards Wavre, keeping them within manageable marching distance of Wellington's army to the west. This was the kind of error that Napoleon would not have made even as recently as the Campaign of France, where the constant flow of intelligence about the precise movement of Coalition forces was essential to French rapid-strike tactics.

After the battle of Ligny, Wellington and Blücher remained in close touch and were well informed of each other's positions, plans and movements. To make matters worse for Napoleon, the French Marshal Emmanuel de Grouchy, who was assigned to shadow the Prussians, lacked initiative and appeared to have received little or no reconnaissance information from his cavalry about Prussian movements.

With large armies operating in a relatively small area, information about their movements was crucial to the outcome, and yet Napoleon was now operating relatively blind. However, he ordered a reconnaissance in force, which allowed him to discover the position taken by Wellington's army, and on the night of 17 June, Napoleon carried out a personal reconnaissance to his outposts.

The next morning, the ground was sodden from overnight rain and Napoleon, having made another personal reconnaissance of the battlefield, compounded his first error of failing to keep close touch with the Prussians with a second, though more understandable one: ordering a delay until the ground dried. While Napoleon waited, Wellington had been busy sending despatches to Marshal Blücher specifically requesting his aid on the battlefield.

At breakfast that morning, Napoleon disparaged Wellington and the British Army in front of his generals, in spite of the fact that Wellington had defeated French armies several times between Lisbon and the Pyrenees. He also underestimated Marshal Blücher's determination to defeat him, whatever the cost.

After the battle of Waterloo had started, insufficient reconnaissance and preparation on the French right flank meant that the Prussians were able to make a substantial advance towards the battlefield before being opposed. By then, Napoleon was throwing everything into the ring against the Anglo-allied army, which was proving to be stubbornly resistant.

The Napoleonic Wars demonstrated that the use of scouts, mainly in the form of cavalry but also increasingly in the form of light infantry, was central to the decision-making process and the positioning and movement of large forces that would win or lose battles.

CHAPTER 4

Beyond the Frontiers

During the 19th century there was plenty of scope for scouting and reconnaissance as peoples and nations explored new frontiers. Whether in India, North America or Africa, this was the period when the scouts pushed beyond the boundaries.

'The Great Game' — Britain and Russia (c. 1830–1895)

While the Napoleonic Wars were mostly focused on Europe, Napoleon also had his eye on Britain's crown jewel, India. Napoleon suggested a joint venture with the Tsar Alexander I, whereby both France and Russia could move into India and share the spoils.

The prospect of invasion became a huge concern for the officials of the British East India Company and political and military administrators. Although Napoleon's ambitions were ultimately thwarted at Waterloo, the threat from Russia remained. It became a priority for the British to assess likely invasion routes for defensive purposes as it was a priority for the Russians for offensive purposes.

The task required a particular kind of exploratory reconnaissance which would be carried out far beyond the frontiers of India, and in areas so remote that there were sometimes no available maps. As Russia sought to extend its influence, it sent out military officers on lengthy probing forays. The expeditions sent by the British and the Russians, exploring and mapping remote areas and carrying out reconnaissance of fortified towns, came to be known as 'the Great Game'. Although the term may have roots in games of risk and deception, such as games of dice or cards, it was more specifically coined by a British political officer, Captain Arthur Conolly when writing to a colleague about Britain's role in Afghanistan and policy towards Russia. The term was also used by Rudyard Kipling in his novel *Kim*, which is set against the backdrop of Anglo-Russian intrigue and rivalry to the north of India.

The main players in the Great Game were ambitious, intelligent and intrepid young officers. Indian hillmen were also sometimes employed to carry out clandestine surveys without attracting undue attention to themselves.

In 1810, Captain Charles Christie, of the Bombay Regiment, and Lieutenant General Sir Henry Pottinger, 1st Baronet, GCB, PC, both officers in the 102nd Prince of Wales's Own Grenadiers, also known as the Bombay Native Infantry, set out on a reconnaissance of Herāt. Taking note of defensive opportunities en route, they reported that the town, while not having particularly impressive fortifications, was set in a fertile valley that could provide plenty of essential provisions for an invading army.

Pottinger then continued on a 900-mile journey that would take him across two extensive deserts. He relied on local guides to lead him to wells and to provide some protection from marauding tribesmen. He kept his compass and notebook concealed.

Despite the personal hardships he experienced on this long and dangerous journey, Pottinger never forgot his purpose of carrying out a detailed reconnaissance of the areas that he passed through. He took surreptitious notes of wells, defensive positions, types of crops and prevailing weather conditions. He also reported on local tribes and whether they were friendly or not. As he advanced on his journey, he created a sketch map which would be used by the military authorities in India as the first map of the area. Among other achievements, he was the first westerner to confirm the position of the Helmand desert. The existence of this and other deserts found by Pottinger and others had important strategic significance for the British as they would constitute major obstacles to an invading army.

Between them, Pottinger and Christie's reconnaissance through areas such as Balochistan totalled over 2,000 miles. They were able to produce detailed accounts of the military and political aspects of the areas that they passed through. Although Pottinger would survive to tell the tale, Christie was killed when Abbas Mirza, Qajar crown prince of Persia, failed to heed Christie's advice about maintaining advanced warning pickets in case of a Russian attack. Sure enough, the Russians did attack, without warning.

On the Russian side, Captain Nikolay Nikolayevich Muravyov-Amursky was sent on a secret reconnaissance mission to Khiva, ostensibly to deliver a letter of friendship to the Khan of Khiva from General Aleksey Petrovich Yermolov, but also to assess the military attributes of the area. As he travelled, like his British contemporaries, Muravyov-Amursky took careful note of the available wells, the defences of Khiva itself and the strength of the Khan's military forces. Muravyov-Amursky not only had a soldier's appreciation of terrain, but he was also a qualified surveyor. Although he was suspected of being a spy on arrival at Khiva, Muravyov-Amursky used his considerable personal charm to negotiate an audience with the Khan and ensure him of General Yermolov's and Russia's good intentions.

Once Muravyov-Amursky had returned to the warship waiting for him on the Caspian Sea, he delivered a detailed report that covered everything from the approaches to Khiva, its local economy, government, arsenals and defences. He also gave his assessment that Khiva could be taken by a Russian force of about 3,000 experienced soldiers. Muravyov-Amursky's reconnaissance would prove to be extremely significant. The information he provided would pave the way for Russian encroachment on the Khan states of Central Asia in the years to come.

William Moorcroft, an English explorer, was head of the horse stud of the East India Company, the immensely powerful trading company founded in 1600 that dominated trade in the Indian Ocean region and beyond for over a century until it came under British Government control in 1858 and which had its own private army. Moorcroft was keen to obtain the exceptionally sturdy Turkoman horses with their pale gold coats. His journeys through Ladakh in 1811 yielded useful information about Central Asia, though it seems that his superiors were more interested in horses than intelligence. He reached Bokhara but succumbed to a fever on his return journey. Although Moorcroft and his companions did not survive their journey to Bokhara, the Russians also had an interest in the city and dispatched a reconnaissance mission to survey the defences and to report on the military and political status of the city.

Captain Arthur Conolly, an officer in the 6th Bengal Light Cavalry, set out on several reconnaissance trips and is credited with first using the term 'Great Game'. Conolly travelled from Moscow to the Caucasus with a Cossack escort in the days when Britain and Russia still had cordial relations. Although he intended to reach Khiva, he was captured on the way and, after his release, headed for Herāt, in Afghanistan, instead. He reached the city in September 1830, and appraised its potential for providing forage for an invading army. Herāt was situated in a fertile valley, and Conolly reported on its strategic advantages. His reconnaissance mission then took him to Kandahar, and from there to Quetta, at the head of the Bolān Pass. This was one of the key potential entry points for an invading army heading towards India.

By the time Conolly had returned to India, he had travelled many of the routes that an invading Russian army would use when approaching India, and this made his reconnaissance particularly valuable to British military planners. The two main routes for a potential Russian invasion that Conolly identified were, first, Khiva, the Hindu Kush to Kabul and the Khyber Pass to Peshawar and, second, Herāt, Kandahar and Quetta to the Bolān Pass.

Conolly's assessment informed his superiors that, apart from the rugged terrain, the Russians would have to contend with the tough Afghan tribesmen, whether they be Pashtun, Turkomen Tajiks or any of the other ethnic peoples, who would harass their movements, cut off their supplies and carry out constant skirmishing from their mountain strongholds. The same would apply to Russian incursions in the 20th century.

Conolly embarked on another expedition in 1841, this time to rescue Lieutenant-Colonel Stoddart, who had been imprisoned by the Emir of Bokhara. However, Conolly was also imprisoned and both he and Stoddart were executed under orders from the Emir who was piqued at not having received a reply to his letter to Queen Victoria. This underlined the real dangers of reconnaissance missions in these remote areas.

Captain Sir Alexander Burnes Kt FRS of the 1st Bombay Light Infantry was also despatched to Afghanistan to gather as much intelligence as he could. In 1831, he explored the route from Kabul to Bokhara and provided detailed assessments of local politics for his superiors. He also travelled up the Indus. One of his reconnaissance missions was on the pretext of delivering a gift composed of horses to the Maharaja Ranjit Singh. Later, Barnes would become a political agent in Kabul, where he was assassinated by an Afghan mob prior to the First Afghan War (1838–42).

Lieutenant Eldred Pottinger carried out a reconnaissance of Herāt in 1837, which was soon besieged by a Persian army with Russian advisers. Pottinger helped the Afghan commander to defend the city against the attacks. He later became a hostage to the Afghan commander Akbar Khan, when the British retreated from Kabul in 1842, and were massacred on their journey through the Khyber Pass.

Major-General Mikhail Chernyaev marched through Turkestan with 1,000 men as Russia sought to extend its influence in the region. Chernyaev excceded his orders by making an assault on the walled city of Tashkent in May 1865. He ordered his scouts to carry out a detailed reconnaissance in which they identified the lowest point in the city wall, against which they could place scaling ladders. They also discovered a secret passage, which enabled them to enter the city while a diversionary attack was taking place on the other side of the city.

Lieutenant-Colonel Thomas Gordon was despatched across the Pamir range of mountains to clarify the intelligence picture in this remote region. When he returned with his findings in the spring of 1874, the results caused consternation as they revealed that both the Broghil and Ishkoman Aghost Passes were vulnerable to attack by a Russian army.

In 1886, Lieutenant Colonel Sir Francis Edward Younghusband, KCSI KCIE of the 1st King's Dragoon Guards set out from Peking, heading west, crossing 1,200 miles of desert. He travelled across Manchuria and the Changbai Mountains before reaching India. He had crossed the Taklamakan Desert and found a way to India through the Muztagh Pass.

Later, he was sent out from Leh, a high-desert city in the Himalayas, into the Karakoram mountains to try to find the Shimshal Pass. During his expedition, Younghusband met the Russian officer Bronislav Grombchevsky, who was also in the area, and the two had frank discussions about their national ambitions over vodka.

During the Russo-Japanese conflict over competing imperial ambitions in Manchuria and Korea, the Japanese launched a surprise attack on the Russian naval base at Port Arthur, in Manchuria, on 8 February 1904. Russia moved its focus away from the Indian frontier to the more immediate threat in the east. More humiliations for Russia followed at the hands of the Japanese at Mukden and Tsushima. The Anglo-Russian Convention of 1907 brought the 'Great Game' to an end, though Britain would later face the threat of a post-Tsarist Russia.

Although Britain, in due course, took control of the Khyber, Karakoram and Bolān Passes as security against Russian attack, they did not control the independent-minded tribesmen who lived there, particularly the Pathans. In view of this, a corps of frontier scouts was created in 1900. Chosen for their hardiness, endurance, initiative, field craft and marksmanship, they were often recruited from the Pathan tribes themselves. Trained and led by British officers and carrying the latest rifles, such as the Martini-Henry, the frontier scouts negotiated the rugged mountainous country, keeping an eye on rebel movements and intervening when necessary with rapid fire and movement.

The American Western Frontier

As the originally English east-coast colonies in New England expanded westwards from the 17th century onwards, their Christian consciences told them that they should respect the integrity of the Native Indian homelands and hunting grounds. However, with the discovery of gold in California in 1848, conscience took a back seat. As the colonists moved further westwards in their wagon trains, they were transformed from European settlers into American pioneers.

The spirit of adventure and endurance exemplified by these pioneers found its apotheosis in the frontier scout. The men who went ahead of the covered wagons to discover and break new ground, forage, interact with, or evade and report on Indian movements, made expansion possible. Many of these scout frontiersmen learned their skills as hunters and trappers and, living among Indians, honed skills in covert movement, attack and evasion. They were also expert trackers, able to strike a trail and read the ground for clues that would enable them to tell how many people or horses had passed that way, at what speed and how recently.

Having grown up learning Indian ways, these frontier scouts had an innate understanding of how the Indians thought and acted, giving them a sixth sense about what they might do. To this was added hawk-like eyesight, keen hearing and extraordinary levels of patience and endurance.

American military commanders were well aware that there was no better way of finding these skills than by employing the Indians themselves. Many of the western scouts were Native Americans or part-Native American. Although scouts occasionally acted alone, military scouting patrols often involved large groups of scouts. From the perspective of the US military, the objective of the scout was to track, to observe and to provide information while also evading the enemy. Sometimes, scouting groups were large enough and armed enough to carry out direct action on small groups of the enemy, but these were the exceptions that proved the rule. Scouting groups would often operate well in advance of a moving column, while also keeping open communications at all times.

Tracking animals, as well as humans, was part of the birth right of Native American Indians and was also learned by American frontiersmen. Mostly taught by word of mouth and example, tracking was a complex subject which required considerable skill. Expert trackers could determine the number of horses or people, the speed at which they were travelling and the sort of burdens they were carrying, along with details such as signs of lame horses or limping people.

Christopher 'Kit' Carson

Although his reputation was to be aggrandised by popular folklore, the facts of Kit Carson's life were extraordinary enough to make him a legend of the Wild West. Starting out as a trapper and pioneer, Kit Carson played a significant role in the expansion of the American frontier. He was employed initially by the explorer John Charles Frémont of the Army Corps of Topographical Engineers, with whom he carried out surveys that took them through the Rocky Mountains, to the mouth of the Columbia river. By March 1844, Frémont and Carson had reached the Sacramento river, having crossed the Sierra Nevada. Carson was later enlisted by General Stephen Watts Kearny, who was charged with nothing less than the acquisition of California for the United States. Carson's knowledge of Indian ways was an important asset, as he negotiated truces and built goodwill as the Americans expanded into what would become the 51st State.

In 1861, Carson raised the New Mexico and Colorado Auxiliary Scouts who would later become the 1st New Mexico Cavalry.

Jim Bridger

Another renowned scout with a similar background to Carson as a fur trader, trapper and mountain man, James Felix Bridger roamed through vast territories often where no white man had ever been. His adventurous probing expeditions took him to the Great Salt Lake in 1824 and to Yellowstone.

Having worked for the Rocky Mountain Fur Company as a trapper and guide, Bridger was enlisted by Captain Howard Stanbury of the US Topographical Corps on the Oregon Trail. His discovery in 1850 of a new route through the mountains, to be named Bridger's Pass, would later by exploited by the Union Pacific Railroad.

In 1857–58, Bridger scouted for Colonel Albert Sidney Johnston in the Utah War, drawing on his unrivalled knowledge of the area and his connections with, and understanding of the Native Americans. His sympathy for Indian culture, and knowledge of their language, including sign language, enabled Bridger to use diplomacy where necessary to avoid confrontation.

All his scouting and diplomatic skills were called upon when he scouted for an expedition of the 18th Infantry under Colonel Henry B. Carrington, with its mission to protect the Bozeman Trail from the Sioux in 1866.

Mariano Medina

Mariano Medina was a contemporary of Kit Carson and accompanied him on the explorations of the west with Lieutenant Frémont in 1842. Medina's scouting skills were employed in the Mormon Rebellion of 1857–58, when he set out with Captain Randolph Marcy from Fort Bridge on 24 November 1857. They waded through deep snow in a featureless wilderness, finally emerging from the forests to reach Fort Massachusetts after an extraordinary feat of scouting and endurance.

Medina set up one of the first stage stations in present-day Colorado, situated at the junction of the various key trails, including the Texas, Overland and old Oregon Trails. When Ute Indians raided his supply of horses, Medina tracked them down for 40 kilometres before shooting one and scattering the rest.

James Pierson Beckwourth

Jim Beckwourth was born a slave with a white father and mixed-race mother. Having joined a fur-trading expedition to the Rocky Mountains, in 1822 he settled for some years among the Crow Indians, which would have enhanced his tracking and scouting skills.

He served as a scout in the Seminole Wars of 1835–43. In 1850, he discovered a pass through the Sierra Nevada, which had a significant impact on opening up California to settlers. In 1884, he served as a scout for the US Army in the Cheyenne War.

Beckwourth was a reluctant scout for the expedition led by Colonel John Chivington, which led to the massacre of the Cheyenne at Sand Creek, on 29 November 1864. This outrage took place despite the fact the Indians were displaying both American and white flags. The massacre led to an extended Arapaho-Cheyenne War, and also contributed to the Plains Wars that followed from the 1850s to the 1870s. This was precisely the sort of incident that scouts, with their knowledge of Native American culture, could have helped to avoid.

Archie McIntosh

Having been of Scottish and Chippewa ancestry, McIntosh was enlisted by General Crook as a scout in the Paiute Wars.

First recruited as an army scout in 1865, McIntosh's skills enabled an army column to escape an attack by Columbia River Indians. His exploits continued into the Apache Wars.

William McKay

Of Scottish and Chinook ancestry, McKay worked with the Warm Springs Scouts, many of whom served in the US Army between 1866 and 1867, during the campaign against the Paiutes. Divided between two units, one of which was under McKay, the Warm Springs Scouts, who had a habit of tracking down their foes by night and then surprising them, developed a fearsome reputation and were regarded as some of the most effective of all US Army Scout units. McKay was also assigned the mission to track down and rescue Native Americans who had been captured by Indians. His brother, Donald McKay, subsequently took command of the Warm Springs Scouts during the Modoc War of 1873. These scouts, armed with Spencer repeating rifles, were highly regarded, both for their scouting skills and their abilities as auxiliary soldiers. However, their fighting techniques were those of the modern elite soldier, operating in twos, with fire and movement and making good use of cover. The Warm Springs Scouts were responsible for tracking down hostile Modoc Indians in a place called Willow Creek where they managed to surround them.

Major George Forsyth

On 24 August 1868, Major George Forsyth was ordered to form a group of 50 frontiersmen to act as scouts. The scouts were armed with a mixture of seven-shot-repeater Spencer carbines and Springfield rifles and all carried Colt revolvers.

In September 1868, the scouts were ordered by General Sheridan to track down marauders who had attacked a freight train along with other raids.

Having left Fort Wallace, they soon struck the trail of the marauders and the scouts made a rapid estimate of numbers. They identified a large body of Indians who were moving slowly, due to the amount of equipment they were carrying. It was clear that the scouts would soon be able to catch up with them.

Having camped on a sand bar on the Republican River, the scouts were attacked by a large force of Indians under the command of

Cheyenne Chief Roman Nose. Digging into the sand and using their dead horses for cover, the scout unit managed to hold off several Indian attacks by accurate shooting with their carbines and by targeting the Indian leaders, including Roman Nose. This had the effect of muddling the Indian chain of command. It was clear, however, that in due course they would either be overwhelmed by sheer numbers or run out of ammunition.

Two of the scouts, Simpson 'Jack' Stilwell and Pierre Trudeau, were tasked with getting through Indian lines and back to Fort Wallace to call for reinforcements. Disguising themselves as best they could with Indian-style blankets, the two scouts travelled by night and hid during the day. On one occasion, they climbed into the carcasses of two dead buffalos to conceal themselves from the Indians. Eventually, they reached Fort Wallace and reinforcements were deployed. Word had also reached Colonel Louis Carpenter, whose men came to Forsyth's aid before the Fort Wallace detachment arrived under Colonel Henry Bankhead. Forsyth himself was wounded while Lieutenant Beecher died of his wounds.

The incident underlined the dangers of scouting operations against a foe possessed of advanced scouting skills themselves. As they had identified a large body of Indians, it might in hindsight have been wise for Forsyth to have sent back for reinforcements earlier and before they were surprised by the Indians.

Native American Scouts

On 28 July 1866, the United States Congress officially authorised the use of Native Americans in the armed forces, with a limit of 1,000. Due to their extraordinary scouting abilities, the Native Americans were almost continually employed as scouts under the command of a white American officer and a civilian chief of scouts.

The Native American scouts soon proved their worth. Their performance on the Yellowstone Expedition of 1871–73 led to the award of a Certificate of Merit.

Black Seminole Scouts

The Black Seminole Scouts, descendants of escaped slaves who sought refuge among the Seminole Indians, were first formed into a unit in Texas in 1870. They proved to be highly effective in their work and demonstrated an extraordinary resilience when following a trail, gradually wearing down their quarry. However, their speed and tenacity also presented a challenge for the soldiers who had to keep up with them.

Placed under the command of Lieutenant John Lapham Bullis, they were officially designated the 'Detachment of Seminole Negro Indian Scouts' and were issued standard military equipment and rifles, although they tended to mix regular uniform with their traditional clothing.

The Black Seminole Scouts were involved in various raids into Mexico on the trail of marauding Indian groups. This led to tension with the Mexican military which was resolved when the Mexican president acted against Indian marauders.

Lieutenant Bullis led the Black Seminole Scouts on an expedition on the trail of Comanche Indians near the Pecos River in April 1875. Having tracked down the Indians, they began a fight from which the scouts had to withdraw. Lieutenant Bullis, whose horse had been shot, was rescued by one of the scouts and taken to safety.

Apache Scouts

During the Indian wars in the North American south-west, the combatants had to negotiate extremely rugged country which tested scouting skills to the maximum. The Apaches who lived in these areas knew the ground like the backs of their hands and most white scouts would be hard pressed to track them down.

Lieutenant-Colonel George Clark of the 3rd Cavalry came up with the obvious solution which was to enlist Apache scouts with equal knowledge and skills. The 3rd Cavalry deployed to Tucson with a detachment of Indian scouts which was designated Company A Indian Scouts, later supplemented by Company B Indian Scouts, though Clark found it difficult to find sufficient numbers of the *crème-de-la-crème* Apaches who were regarded as the most effective scouts in these areas. The Apache scouts soon proved their worth, tracking down renegade Apaches in the mountains with remorseless efficiency.

Black Hills Expedition

Led by Lt-Col George Armstrong Custer of the 7th Cavalry, the Black Hills Expedition set out for an armed reconnaissance of the previously uncharted Black Hills of South Dakota with a view to building a fort in the area as well as exploring new routes through the mountains with the potential for gold mining. The expedition included several scouts led by Bloody Knife.

Custer reported that there was evidence of metals in the rivers which, when telegraphed to the east, led to a gold rush into the area. This in turn increased tensions with the local Sioux Indians.

Southern Plains War

Frustrated by the decimation of buffalo herds which were the basis of their livelihood and bothered by the continual intrusions of white settlers into their nomadic way of life, Plains Indians, including Cheyenne, Comanche and Kiowa, became more aggressive towards the interlopers.

General Philip Sheridan ordered no fewer than five army columns to move towards the area as part of a strategy to wear down Indian resistance.

The Indians had first to be located and this was the duty of the scout units that accompanied the army detachments. The Indian scouts attached to the army tracked down the Indian groups who were then forced to move on, leaving their possessions and food behind.

In one battle, at Palo Duro Canyon, Black Seminole Scouts reported the location of a large Indian village consisting of Comanche, Kiowa and Cheyenne. On this occasion the Indians were caught by surprise by the US Army troopers and did not have time to escape with their horses, possessions and stores of buffalo meat. Bereft of food and space in which to carry on their way of life, many Indians had no option but accept containment in the reservations while the land on which they had roamed for generations was taken over by white American farmers and ranchers.

The Powder River Expedition and the Battle of the Little Bighorn

Led by the charismatic Lakota Chief Sitting Bull, the Plains Indians who had resisted confinement in reservations came out in open rebellion. Inspired by his leadership, many Indians who had moved to the reservations decided to come out again in order to join the free spirits for the spring 1876 hunting season. By June 1876, a large number of Indians had moved to a camp on the Little Bighorn River in Southern Montana.

Lt.-Gen. Philip Sheridan ordered several army columns to converge on the area occupied by the Lakota in an effort to contain and subdue them. The army column set out from Fort Ellis under Colonel John Gibbon, from Fort Fetterman under General George Crook and from Fort Abraham Lincoln under General Alfred H. Terry, including the 7th Cavalry under General Armstrong Custer.

General Terry's plan was for Custer to approach from the south, pushing the Indians northward, where Terry would cut them off with another force further upriver.

Custer had with him some notable white scouts, including Charlie Reynolds, who had served with Custer on the Black Hills expedition, George Herendeen and Billy Johnson. There was also a group of Native American scouts, including Bloody Knife, who was much admired by Custer.

On 22 June, Terry ordered Custer to carry out a reconnaissance in force along the Rosebud river, giving Custer some leeway to use his initiative. During their reconnaissance of the area, the scouts, having climbed to a knoll that separated the Little Bighorn and Rosebud rivers, identified a substantial Indian village. The obvious signs were

General Custer with some of his scouts. (Wikimedia Commons)

a large herd of ponies and several teepees. However, when an officer and General Custer were called to take a look for themselves, they claimed that they could not see anything. Whether this was because the scouts had better eyesight than the officers is not clear.

Planning to make an attack the following morning, Custer was then informed that hostile Indian scouts had identified the trail of the 7th Cavalry and that their position was therefore known to the enemy. Custer now decided that there was no time to lose and that he must order an attack immediately.

Custer's scouts tried in vain to convince him that he was hopelessly outnumbered. Bloody Knife was so exasperated by his commander's intransigence that he stripped off his military uniform saying that

Bloody Knife was Custer's favourite scout. He was killed as the battle began. (National Archives)

he would prefer to die as a Native American. Another scout, Mitch Bouyer, told Custer that this was the largest Indian gathering that he had encountered in 30 years.

Ignoring the warnings, Custer decided to divide his command in three, despite having been told that even his unified command was greatly outnumbered. One battalion was placed under the command of Major Marcus Reno; another under Captain Frederick Benteen while the third was under his own command. By midday, Custer's divided command set out towards the Indian camp and its date with destiny. Custer's Indian scouts could do little more than look on as the force that it was their duty to protect moved towards a fate that they had accurately predicted.

Custer partly based his decisions on the estimates that had been received by the military from Indian agents who had kept a tally of the numbers of Indians in reservations and those outside the reservations. The recorded number for those outside reservations was modest due to the failure to take into account the significant numbers of Indians who had left the reservations secretly. Trusting in the estimates of men with clipboards, Custer ignored the advice of the scouts who had assessed the situation before them in real time.

Custer's plan appears to have been to take Indian women and children hostage in order to persuade the braves to comply with the authorities. However, having been advised by his Crow scouts that the village had been alerted to his presence, Custer ordered Major Reno to attack. As Reno's companies advanced, they were shielded from view of the village and could not themselves appreciate the size of the village until they were a few hundred yards from it, by which time it was too late. At this point, Reno realised what he was up against and ordered his companies to halt, dismount and form an all-round defence. This was quite different from Custer's plan that Reno would drive the Indians scattering before him into Custer's waiting arms. Reno's men fired into the village, stirring up a hornet's nest of braves who then immediately attacked his position. Focusing on Reno's exposed left flank, the Indians pushed Reno's companies back into the woods near the river. Reno then attempted to escape across the river, harassed by Cheyenne Indians. The scout Bloody Knife was killed by a bullet to the head just before the mayhem began.

As they retreated up the bluffs on the river, Reno's men were fortunate to be reinforced by Benteen's companies who happened to be passing through. This helped to prevent the almost certain annihilation of Reno's battalion.

Meanwhile, about four miles to the north, Custer, having manoeuvred to provide a cut-off force, was heavily engaged by an overwhelming force of Lakota and Northern Cheyenne Indians. There are different accounts of Custer's movements since none of his battalion survived to tell the tale. He may have attempted to cross the river to engage with the Indians who he assumed would be fleeing in his direction after Reno's advance but was forced to retreat himself by an overwhelming Indian force.

William 'Buffalo Bill' Cody

William Cody was one of the most famous scouts of the Plains Wars. He worked for the Russell, Majors and Waddell freight company and he is also thought to have been a rider for the Pony Express. One of his riding exploits involved retrieving horses that the Sioux had stolen from a trading station.

Cody worked as a scout for the Union during the American Civil War. This involved him in shooting forays against Cheyenne and Kiowa Indians. He then enlisted in the 7th Kansas Cavalry for the duration of the war and as a civilian scout. He would gain his name as a buffalo hunter to supply workers at the Union Public Railroad, personally shooting over 4,000 head.

Cody's reputation continued to grow as an army scout, drawing on his remarkable knowledge and memory of terrain, endurance and courage. He possessed all the patient qualities of a scout along with a large dose of bravado. He worked with the US 5th Cavalry west of the Mississippi to combat continued Indian resistance to settlement. For his work tracking down stolen army horses for the 3rd Cavalry near Fort Macpherson in Nebraska he was awarded a Congressional Medal of Honour.

Cody participated enthusiastically in the reckoning after the disastrous defeat of Custer's force at Little Bighorn and had a one-to-one duel with the Cheyenne warrior Yellow Hair.

Wounded Knee, 1890

Worn down by the inexorable advance of white settlers and US military forces, the Plains Indians sought spiritual inspiration and protection in rituals such as ghost dances which were designed to invoke the spirits of the past. They believed these spirits would defend them from the attacks of the white man. These dances were regarded by US authorities as war dances and the prelude to an Indian uprising. They therefore decided to intervene.

Ironically, it was Cheyenne and Crow scouts enlisted in the 1st Cavalry who brought news of the growing ghost dance movement to the authorities. As a result, US army regiments were sent out to

The frontiersman and scout Buffalo Bill Cody was a legend in his own lifetime. (Wikimedia Commons)

subdue the Indians, some expeditions accompanied by scouts such as Buffalo Bill Cody.

During one of these expeditions, the Lakota Sioux Chief Sitting Bull was killed. On 28 December 1890, a detachment of US cavalry escorted Lakota Indians to Wounded Knee creek where they were joined by a further detachment of cavalry under the command of Colonel James Forsyth, who were armed with rapid-fire Hotchkiss guns.

Oglala scouts were sent in to disarm the Lakota which led to a scuffle and a weapon being fired. This was followed by an exchange of fire that left up to 300 Lakota dead, including 60 women and children. About 30 US military personnel were killed.

Elsewhere, other Indian scouts successfully persuaded Indian ghost dancers to return to the reservations, though on one occasion the scout commander Lieutenant Edward Casey of the 22nd Infantry was killed by an Indian warrior.

Southern Africa

Frederick Russell Burnham

Frederick Russell Burnham was born in a Dakota Sioux Indian reservation where he grew up learning Indian ways. A tough and independent youth, he also learned from the frontiersmen of the old west. One of these was a man named Holmes who had served with John C. Frémont, Kit Carson and other legendary scouts. In the mountains of Arizona and New Mexico, Holmes taught Burnham the skills of scouting, including trailing and hunting, how to double back and cover tracks, direction-finding, time and distance at night and the practicalities of survival such as finding water.

Burnham also learned the importance of understanding the ways and customs of the people he was working among. Part of scouting is understanding what people are likely to do and how they think. Like the Sioux warriors who were hardened by training, Burnham also learned how to endure both in mind and body, for a scout ultimately has to rely on himself rather than the support of a group. Burnham

Frederick Russel Burnham, having just received the Distinguished Service Order (DSO) from Queen Victoria. (Library of Congress)

also learned the art of catching an hour or so of sleep at will, for a scout cannot necessarily rely on regular sleeping hours. His five senses were sharpened by constant use in the natural environment and even more so by danger. The scout has to learn to be alone for long periods and endure without familiar human comforts such as shelter, warmth, food and running water.

Burnham's experiences in Apache country gave him a sixth sense with regard to covering ground, whether walking, running or riding. He knew that even a horse in mountainous country cannot outrun an Apache and when on foot he developed a similar fast pace that would keep him ahead of any pursuer.

The American west having been won, Burnham dreamed about another frontier, in Africa. Inspired by Cecil Rhodes, Burnham was convinced that his hard-earned scouting skills from the American west could be put to good use in the service of the pioneering Rhodes.

First Matabele War (1893–1894)

As Rhodes' British South Africa Company moved north in the quest for more land, it ran into opposition from the Matabele under King Lobengula. It was thought that the conflict could be resolved quickly by capturing Lobengula and Burnham and other scouts were sent forward to assess the possibilities.

Having been tasked with finding the Matabele capital, Bulawayo, Burnham and his companions set out into the unknown country, stopping every few minutes to register the route they had taken, knowing that they might need to backtrack in a hurry if they were surprised by the enemy. When they halted, they also observed their horses closely for any signs of twitching ears and sensitive nostrils that might indicate the presence of an enemy.

Having been on one occasion surrounded by native warriors, Burnham and his companion Viverson made their way back towards the column in the dark. At one point, Viverson produced his compass and said that they were heading north, which would take them back towards the enemy. Burnham, however, trusted in his own judgment and memory pictures and insisted that they should stay on course. Later they discovered that Viverson's compass was broken.

There were many other scouts on the expedition, including Hottentots, used to hunting antelope, Boers of the Transvaal, Australians and other frontiersmen.

Burnham and his scouts continued with their work, keeping an eye out to report the presence of the enemy and finding places for the column to halt that were not susceptible to enemy attack. Burnham would set out during the night at around 10 o'clock to recce the route for the following day, returning to the column before dawn.

In the Second Matabele War, Burnham and Bonar Armstrong were tasked with tracking down the Matabele spiritual leader who had done much to foment the uprising. They found their way to the spiritual leader's cave and had to evade a village of Matabele, leaving their horses and crawling along the ground with bits of brush as camouflage. They reached the cave where they shot the leader and then made their escape, followed by hordes of warriors but setting fire to village huts in order to cause confusion. Having removed the influence of their spiritual leader, Rhodes managed to make peace with the rebels.

In the Second Boer War, Burnham was up against a different kind of foe – the Afrikaans-speaking farming settlers of the Cape. These were extraordinarily tough and independent-minded people. Serving under Field Marshal Frederick Roberts, Burnham was appointed chief of scouts, an unusual honour for an American citizen serving with British forces. Burnham was recommended for the post by General Frederick Carrington who described Burnham as the finest scout in Africa.

Burnham was sent out on several dangerous scouting forays against a formidable enemy who deployed their own screen of scouts. Before the battle of Petersburg, Burnham was ordered to enter the town and find out how many Boers were present. On this occasion he could not get through the Boer scout screen and suffered the humiliation of reporting his failure to Lord Roberts. However, the commander in chief never lost faith in his chief scout and Burnham got through more times than not.

On another occasion Burnham hid in a hut which was then occupied by some Boer commanders. Fortunately, they did not notice him under some blankets in a corner. Later Burnham was captured when signalling to British forces that they were walking into a trap set by a Boer commando. He pretended to be wounded so that he could be placed on the wagon for the wounded from which he later made his escape by sliding down between the wheels of the wagon and letting it roll over him.

Robert Baden-Powell

Although rightly celebrated as the founder of the Boy Scouts movement, Robert Baden-Powell's interest in scouting was forged in a military context during service in the British Army in both India and Africa.

Having served in the 13th Hussars in India, Baden-Powell then moved to Africa in the 1880s where he developed scouting skills among the Zulus in Natal province. After a spell as an intelligence officer in Malta, he returned to Africa in 1886 for the Second Matabele War. During operations around the siege of Bulawayo, he commanded reconnaissance expeditions in the Matopos hills and it was in this period that he also met Frederick Russel Burnham, who enhanced his scouting knowledge with his knowledge of woodcraft from the Old West.

Baden-Powell returned to India where he commanded the 5th Dragoon Guards, though he did not forget his interest in scouting. His manual *Aids to Scouting* was intended to introduce recruits to the art and it included some of the wisdom he had learned from Burnham.

He returned to Africa before the beginning of the Second Boer War and soon became involved in the siege of Mafeking, which made him a household name and national hero. Although Baden-Powell used considerable ingenuity in resisting the siege by the Boer Army, including ruses to make the British garrison seem larger than it was, there was some doubt as to whether the siege should have been allowed to happen in the first place, given that Baden-Powell's orders had been to maintain a mobile patrol in the area and not a fixed garrison. During the siege, Lord Edward Cecil came up with the idea of recruiting boys to perform tasks such as delivering messages, duties which they performed with considerable pluck. Baden-Powell paid tribute to them in the introduction of his book *Scouting for Boys*:

> We had a wonderful example of how useful Boy Scouts can be on active service, when a corps of boys was formed in the defence of Mafeking, 1899–1900.

In 1903, Baden-Powell became Inspector-General of Cavalry where he continued to develop ideas for reconnaissance with cavalry, moving away from traditional ideas about the use of mass cavalry. All these ideas would later be obviated by the advent of mechanisation.

In his military guide *Reconnaissance and Scouting*, Baden-Powell places considerable emphasis on gathering information with the use of accurate sketch maps and drawings. There was no doubt in Baden-Powell's mind as to the value of good reconnaissance:

> SUCCESS in modern warfare depends on accurate knowledge of the enemy, and of the country in which the war is carried on.

His views on the role of the scout as the eyes and ears of those in command are also underlined:

> Scouts are the eyes and ears of an army, and on their intelligence and smartness mainly depends the success of all operations. The brain and strong arm, the General and his troops, are helpless unless the scouts explain where, when, and how to strike to ward off attack.

This is further emphasised in his definition of reconnaissance:

> Definition – Reconnaissance is the acquisition of knowledge of the country over which military operations are likely to be carried on, and of the numbers, position, and probably intentions of the enemy; and the better the officer commanding the force knows these particulars, the better he will be able to make his dispositions for attack and defence.

The individual qualities of the scout, including initiative and determination are described by Baden-Powell, including the scout's ability to handle a horse:

> Their Qualifications – A scout must be a man of intelligence and pluck, and a good horseman, with confidence in himself, that is to say, one who will not lose his head in a sudden emergency, but can trust himself to get out of all difficulties, and who is full of "dodges" to meet every kind of incident or accident that may occur.
>
> A man of this kind, enjoying good eyesight and power of hearing, is personally fit to be a scout, but he still requires to be specially trained and, above all, instructed as to what sort of information he is required to obtain, and in what recognised form it is to be recorded or sent in.

The cover of *Scouting for Boys* by Robert Baden-Powell. The note from Baden-Powell underpins the Scout ethos of duty and service. (Wikimedia Commons)

His instructions are very much of their time and he does not seem to take account of the lethal effectiveness of an enemy sniper against a scout:

> Get as near his position as you can without being under close fire – say 500 yards. Select a base line running parallel to the enemy's line, then take the angles of the different chief points of his position from both ends of your base, or else proceed along the base, noting down the different points as you come opposite to them, and taking their distance from you by the method I gave you for ascertaining the width of a river; but the first would be almost always the best and quickest way. When you have fixed the chief points of the position itself, sketch in leading up to the position – and you must stick at nothing in gaining this sort of information: ride about at a trot as close as you can to the enemy, taking advantage of cover and following it along to see how near the position your own side could

approach under it. The enemy may fire at you if they are in a defensive position, but remember it takes an average of a man's weight in bullets to kill him, and that is 1400 rounds, there is such missing in action, especially if the object keeps moving; and if the enemy are in position as outposts they will not fire at you, as it would alarm the whole of their line; they would only send out a patrol to drive you away, which it should never succeed in doing altogether; you may ride away from that part of the line and come back somewhere else a little farther along.

His comments on the observational awareness of the scout echo those of Burnham who kept memory maps in his mind of ground he had travelled over, to the extent that he was confident enough to override the evidence of a compass:

Nothing should ever escape the eye of a scout; he should have eyes at the back of his head: he should take a pleasure in noticing little trifles or distant objects that have not struck the attention of his comrades. Always notice all peculiar features and landmarks while going over strange ground, especially by frequently looking backwards so that you may be able to find your way back again by them. A scout who loses his way is utterly useless.

Remember after passing a difficult place, such as a deep ravine, high fence, thick patch of underwood, &c., to look back at it so that you will be able to recognise the place of passage again at once should you want to come back that way in a hurry, pursued by the enemy, &c.

We can also see echoes of Burnham in his advice to memorise by day and move at night:

Suppose you have found the enemy's outposts, but they are too vigilant or well posted to allow you to get within their line for further information: go along their front and notice any streams, hollow roads, fences, &c., along which you could creep at night, and so get into their position. Select several, and notice particular features connected with them by which you will be able to make use of them at night.

Other tips include how a scout should learn how to orientate himself without a map or compass, such as knowing that a church 'generally points east and west, with the great window at the east end, and tower at the west end.'

His advice on the use of horses in scouting forays includes trusting the horse's instincts when riding over rough ground or jumping difficult obstacles.

Frederick Courteney Selous, DSO

Born in 1851 into an aristocratic family, Frederick Courteney Selous showed an interest in the outdoors and wildlife from an early age. When found sleeping on the floor at the age of ten, he is said to have announced that he was hardening himself in order to become a game hunter in Africa. This is exactly what he would become.

Aged 19, he travelled to South Africa and then on to Matabeleland, where he was given unlimited permission to hunt game by Lobengula Khumalo (1845–1894), the second and last king of the Northern

Frederick Courtney Selous, hunter-turned-conservationist, made his home in the vast plains of Africa. (Library of Congress)

Matabele people. While shooting game and sending specimens to the natural history department of the British Museum, Selous travelled over huge distances and became very familiar with the land and its people. When in 1890 he joined Cecil Rhodes' British South Africa Company, his knowledge and experience proved invaluable and he acted as a guide to the pioneering expedition into Matabeleland. In 1893, when the British South Africa Company under Cecil Rhodes sought to settle the area, Selous took part in the First Matabele War and was wounded during the advance on Bulawayo, during which he formed part of the scouting expedition. At this time, he met Frederick Russell Burnham.

Selous played a prominent part in the Second Matabele War (1896–1897), serving in the Bulawayo Field Force. He met Robert Baden-Powell during this period, who had joined the staff of the British Army headquarters in Matabeleland. The scouts of the Bulawayo Field Force were under the command of Captain George Grey, and Selous and Burnham became involved in many close scrapes with the rebels. As Selous recalls:

> Captain Grey sent the American scout Burnham, together with a companion named Blick, to the top of a hill ahead, to try and ascertain the numbers and disposition of the rebels; but Burnham and his companion were cut off from the main body and had to gallop for their lives.

Selous also got into a tight spot when his pony took fright at an inopportune moment when he was dismounted, taking aim at approaching rebels:

> In front of me lay a piece of perfectly open ground extending along the Umguza, some 200 yards broad, whilst from the edge of the open to the left the country was undulating and very scantily covered with low bush. The pony I was riding was the same that had been lent to me on the previous Sunday, and he had proved himself so absolutely steady, with rifles going off all round him, and bullets pinging and buzzing past him, that the last thing I thought of was that he might now play me false and run away. However, this is what happened. I had dismounted and was sitting down to get a steady shot when someone said close behind me, "Look out, they're coming down on us from the left." I did not know that

any one was near me, but on getting up and looking round, saw one of the officers of the Colonial Boys – now Captain, then Lieutenant Windley – close behind me. At the same time I saw Grey's Scouts retreating on the other side of the river, and recognised that Windley and I were a long way ahead of John Grootboom and five or six other Xosa Kafirs, who were the only members of the corps I could see, and who were also retiring; whilst I also saw that some of the Matabele we had been chasing had rallied, and seeing two white men alone, were coming down on us as hard as they could, with the evident intention of cutting off our retreat. However, they were still some 250 yards from us, and could I but have mounted my pony, we could have galloped away from them and re-joined the Colonial Boys easily enough. A few bullets were again beginning to ping past us, so I did not want to lose any time, but before I could take my pony by the bridle, he suddenly threw up his head, and spinning round trotted off, luckily running in the direction from which we had come. Being so very steady a pony, I imagine that a bullet must have grazed him and startled him into playing me this sorry trick at such a very inconvenient moment. "Come on as hard as you can, and I'll catch your horse and bring him back to you," said Windley, and started off after the faithless steed. But the brute would not allow himself to be caught, and when his pursuer approached him, broke from a trot into a gallop, and finally showed a clean pair of heels. When my pony went off with Windley after him, leaving me, comparatively speaking, plante le, the Kafirs thought they had got me, and commenced to shout out encouragingly to one another and also to make a kind of hissing noise, like the word 'jee' long drawn out. All this time, I was running as hard as I could after Windley and my runaway horse. As I ran carrying my rifle at the trail, I felt in my bandoleer with my left hand to see how many cartridges were still at my disposal, and found that I had fired away all but two of the 30 I had come out with, one being left in the belt and the other in my rifle. Glancing round, I saw that the foremost Kafirs were gaining on me fast, though had this incident occurred in 1876 instead of 1896, with the start I had got I would have run away from any of them.

After the war, Selous returned to his game hunting, acting as a guide for, among others, President Theodore Roosevelt. During a trip with Roosevelt in 1909, Selous recognised that the days of unrestricted big game hunting were numbered, and he began to take measures to create game conservation areas to preserve the diversity of African wildlife for future generations. The poacher had turned gamekeeper.

With the outbreak of war against Germany in 1914, Selous was keen to offer his services, though it was a struggle to get enlisted due to his age. He took a commission with the 25th Royal Fusiliers (the Legion of Frontiersmen) in spring 1915, and by the summer he had been made a captain. In September 1916 he was awarded the Distinguished Service Order (DSO). On 4 January 1917, Selous was leading his company in an operation near Kisaki by the Rufiji River in Tanganyika. Carrying out a personal reconnaissance, he raised his head to look through his binoculars and was shot through the head by a German sniper.

Although he had been persuaded to take part in the expedition on behalf of the British South Africa Company, Selous was sympathetic to the plight of native tribespeople who had been mistreated by some white settlers. He also had sympathy for the Boers and opposed the Jameson Raid. Although married to an English woman, Selous had a long-term relationship with a woman from one of the African ethnic groups. In short, despite his conventional public-school upbringing, his membership of the Royal Geographical Society and his popularity at home and abroad, Selous was ultimately his own man. Like Burnham and Baden-Powell, he was a born scout and died as a scout. The Rhodesian armed forces would recognise this later, when they formed the elite special operations force called the Selous Scouts.

The artist and travel writer J. G. Millais said of Selous:

> If there was one striking feature in his physiognomy it was his wonderful eyes, as clear and as blue as the summer sea. … They were the eyes of the man who looks into the beyond vast spaces. Instinctively, one saw in them the hunter and the man of wide views.

The Lovat Scouts (1900–)

When Simon Fraser, 14th Lord Lovat, began to recruit ghillies, deer stalkers and shepherds from the highlands of Scotland, he was convinced that they would make excellent material for military

Lovat Scouts in camouflage. (Wikimedia Commons)

scouting duties. Their subsequent performance during the Boer War in South Africa and in two world wars would prove him right.

One of the reasons why the British army had such a parlous experience at the hands of the Boers, was that they hardly ever knew where the Boers were located until it was too late. The Boers were well in advance of the British in fieldcraft, reconnaissance and marksmanship. With the arrival of the Lovat Scouts, highly mobile either on tough Highland ponies or on foot, British commanders at last had eyes and ears on the ground giving them warning of potential attacks and also the ability to catch the Boers by surprise.

Not only were the Lovat Scouts skilled in covert movement, they were also equipped with powerful telescopes that allowed them to see well beyond the range of conventional binoculars. They were also trained in signalling either by semaphore, heliograph or, if necessary, by using carrier pigeons.

First commanded by Captain, the Honourable Edward Oliphant Murray, of the 1st Cameron Highlanders, the Lovat Scouts had officers and non-commissioned officers drawn from several distinguished regiments including the Life Guards, Scots Guards, Grenadier Guards, 1st Argyll and Sutherland Highlanders and Northamptonshire Regiment.

The Boers, masters of the ground, now had to contend with a substantial body of men who understood fieldcraft as well as they did. Lieutenant E. Fraser-Tyler describes an incident that demonstrates the transfer of stalking skills to military use:

> Captain Macdonald and Dugald Macdonald stalked a Boer sentry over bare ground, till within 500 yards, and had the light not suddenly failed, they would have got right in and killed him. They did, just as one would with a close stalk after deer. That is, they crawled forward a few yards, while the other kept his telescope on the sentry, and so they crawled closer and closer, stopping whenever the sentry looked their way.

At Vedefort, a Lovat Scout observation post (OP) spotted a large Boer concentration of wagons and men and heliographed a report to headquarters (HQ). No action was taken on this occasion, but later analysis based on information gathered at the end of the war showed that if the British had acted on the information they had received from the scouts, they could have achieved a significant victory over the Boers.

Near the Rodebergen mountains, General Archibald Hunter sent out Lovat Scouts to ascertain the position of the enemy. The mountains had numerous hiding places for Boers, and the scouts were sent out on forays, either individually or in groups, to spy the land and spot any enemy movements. General Hunter knew that he could always count on the accuracy of their reconnaissance reports.

On another occasion, General Macdonald ordered three Lovat Scouts to recce Vaal Krantz ridge at night to see if an attack could be made on the Boer positions in the valley beyond. The scouts climbed to the summit of the snow-covered ridge on a cold, wet night and found the ridge itself was deserted. They could see the Boer campfires in the valley below. The scouts moved rapidly back to the British lines and a Highland Light Infantry company was despatched immediately guided by three scouts. They encountered the Boers who were returning to the ridge and drove them back. The Boers were forced to retreat from Retief's Nek.

The Tigers

The Tigers were a force of 200 colonial horsemen raised from English-speaking South Africans who fought in the Second Boer War (1899–1902). It was also a requirement that they should speak the Dutch settler-derived language Afrikaans and at least one indigenous Bantu language. They were initially commanded by Major, later Colonel, Mike Remington, and their nickname was derived from the big-cat skin, in fact from a leopard, which was wound round their slouch hats. These scouts were issued with pistols and carbines and rode out ahead of the main force. A *New York Times* war correspondent wrote admiringly:

> These magnificent rough riders are all well mounted, good shots, and keen of sight. They ride light, and, as the they scour the country before their heavy army, not a beast or a human being, not a suspicious rock, or a dangerous mountain pass, escapes their attention. Many of the South African colonial scouts know the country like the palms of their hands: their instinct tells them where to look for the enemy, and how to take him unawares. These men will go out for weeks at a time on scouting expeditions and will think nothing of doing 60 or 80 miles in the 24 hours. They can sleep, as can all good scouts, at any moment, awaking at the time desired; but they will not be caught napping.'

Boer Scouts

Although the Boers were natural countrymen and sharpshooters, General Christiaan Rudolf de Wet (1854–1922) relied upon a specialised reconnaissance force to give him the vital information he needed about British Movements. One of these scout units was led by

Daniël Johannes Stephanus 'Danie' Theron. This scout unit, known as the *Theron se Verkenningskorps (TVK)* (Theron's Reconnaissance Corps), was equipped with bicycles. In regard to his appearance, Theron did not immediately fit the bill of the Boer frontiersman. He was a neatly turned out schoolmaster with an irascible temper. However, he proved to be prodigiously effective as a scout.

The *TVK* made notable contributions at the battles of Sion Kop (23–24 January 1900) and at Paardeberg (18–27 February 1900), where Theron personally made his way twice through British lines.

The *TVK* were equipped with binoculars, revolvers and carbines and they were constantly on the watch for British movements, while also studying British tactics during battles. Apart from their reconnaissance missions, the *TVK* also blew up bridges and railway lines.

The *TVK* was so effective, that the Commander-in-Chief of British forces, General Lord Roberts, put a large reward on Danie Theron's head and sent out 4,000 British soldiers to eliminate the force.

Danie Theron was killed by chance when he ran into a group of South African cavalry called Marshall's Horse (1899–1902). Having killed or maimed several of the riders, Theron was killed himself by an exploding shell fired by British artillery who had been alerted by the shooting.

Gideon Scheepers (1878–1902)

Gideon Scheepers was an experienced heliographer in the Boer State Artillery and was later recruited by General Christiaan Rudolf de Wet as captain in charge of his reconnaissance corps. He took part in commando raids against the British, which included sabotage of railway lines and telegraph lines.

He was captured by the British and later executed for war crimes, which included shooting native scouts who had been sent out by the British, and who Scheepers decided were spies.

CHAPTER 5

The First World War

The First World War presented unique challenges so far as scouting and reconnaissance are concerned due to the particular nature of the entrenched warfare. Those tasked with reconnaissance missions had to call upon high levels of stealth and courage. Reconnaissance was sometimes associated with sniping skills as well as with the work of artillery observers.

In August 1914, the plan based upon the ideas of German General Alfred von Schlieffen (1833–1914) began to unfold. Powerful German armies advanced in what was planned to be a great wheel that would take them past the French western fortifications, through Belgium and the south Netherlands and round to the north of Paris. Under its misguided Plan XVII, by which French forces would plunge through Alsace-Lorraine into German industrial areas, French planners radically underestimated the size of German forces that would be deployed against them and also unwittingly strengthened the potential for encirclement by German forces in the Schlieffen Plan. Having taken the Belgian forces in Liége, the Germans continued towards the Meuse and Lorraine where their own plans began to fragment as commanders such as Moltke made contrary decisions. As the Schlieffen Plan began to unravel, General Joseph Jacques Césaire Joffre wisely dropped Plan XVII, and devised a new plan, as it were, on the back of an envelope, with

French defences pivoting on Verdun. German forces continued to head towards the Marne, but the French were able to defend Paris.

During this period of considerable movement of large forces, cavalry continued to be a major source of tactical reconnaissance. From August 1914, the Germans carried out tactical reconnaissance at a divisional level with a light cavalry squadron, while a light cavalry regiment was deployed at corps level. The French deployed cavalry in front of their advancing infantry, though sometimes it was so far in advance, it was out of useful touch. On occasion, French and German armies ran into each other with little or no warning.

As the movements of forces slowed down and became entrenched, and as both light and heavy machine guns were deployed in large numbers, cavalry ceased to be a viable arm for either reconnaissance or attack. Aerial reconnaissance continued to grow in importance, while from September the opposing armies deployed infantry for reconnaissance according to their own capabilities and requirements. In the somewhat novel circumstances of trench warfare, the decline of cavalry left a vacuum that would be filled by a mixture of static observation posts, sniping activity and mobile observers and patrols.

Scouts, Observers and Snipers

The manner of Frederick Courtney Selous' death was significant, for it underlined the growing importance of the sniper, which in the static warfare of the trenches of Western Europe between 1914 and 1918, as well as in Gallipoli between February 1915 and January 1916, would wield a reign of terror. It was also the case that, at the beginning of the war, at least, the German army was better qualified and better equipped at sniping and associated reconnaissance than the British.

Men chosen for reconnaissance and sniping duties often had a background in country activities such as gamekeeping, stalking and hunting. They had an instinct for fieldcraft and for camouflage.

One such was the poet Julian Henry Francis Grenfell, an officer in the Royal Dragoons. Grenfell had volunteered more than once for sorties beyond the British lines, where he made good use of the stalking skills he had learned on the Scottish moors. Moving in daylight to within metres of the enemy trenches, Grenfell shot three Germans on separate occasions, and in November 1914 also observed preparations for an imminent enemy attack. He moved quickly back to his own lines to alert his unit and was awarded a DSO for his 'excellent reconnaissance', daring and ingenuity. Despite the danger to which Grenfell exposed himself in no man's land on his sharpshooting and reconnaissance forays, he was killed by a shell fragment in May 1915 while chatting to fellow soldiers in the relative safety of British lines.

The Germans began the First World War with a large number of well-trained and well-equipped snipers. Their weapon of choice was a specially adapted version of the Gewehr 98 rifle, of which at least 18,000 were produced to be fitted with telescopic sights made by a variety of leading German manufacturers, including Zeiss, Gorz, Hensoldt and Voigtländer. The Gewehr 98 had an effective range of 800 metres. More importantly, the German snipers, largely recruited from among gamekeepers and stalkers, were well trained and knew how to move covertly into optimum positions and, also, how to operate and care for the sophisticated optics in the telescopic sights.

By contrast, the British were well behind. In the early stages of the war from 1914, barely trained sharpshooters were firing over iron sights and, when issued with telescopic sights, were not trained how to adjust them. British regiments were losing about five men per day to German snipers. Realising that the British needed to catch up fast, Major Hesketh Vernon Prichard, who had initially been sent to France to take charge of war correspondents, recommended the establishment of a Scout, Observation and Sniping School to provide training in areas such as fieldcraft, weapon handling, camouflage, observation and report making. The British realised that the value of a sniper lay not merely in acquiring targets, but also in what he could see in and behind enemy lines. They maximised this observation role by sending out snipers with an observer, whose duty was to acquire targets for the sniper and to provide observation reports. These would include the time and grid reference for particular sightings, and this information would be passed on to intelligence. This might range from the identification of a German cap badge, and thereby the unit in that area, to the digging of new earthworks or the site of a machine-gun nest.

The Lovat Scouts (Sharpshooters), who were formed by Lord Lovat in the autumn of 1916, proved to be the most adept at these tasks but, although they were officially designated as sharpshooters, their real value lay in observation, and it was for this role that they were invariably called upon. Like the telescopic sight, an observation telescope required experience to be handled effectively and, for the Highlanders, this was in their blood. Observers such as the Lovat Scouts or Corps Cavalry observers would not only observe enemy lines when they were relatively static, but also report on developments during battle, such as when a particular unit had reached a particular objective or enemy movements. These observations were relayed back to high command, enabling them to make appropriate plans. Battle observers would often start in observation posts in their own lines and then move forward as the battle developed to appropriate places of concealment in the advanced shell lines. They would use telephones for as long as was possible and if the lines were cut by shellfire, they would send runners with their latest reports.

The Lovat Scouts, among others, favoured Ross telescopes, which they were familiar with from their stalking days in the Highlands. From the 1890s, Ross collaborated with German lens makers such as Carl Zeiss and Carl Paul Goerz. The British Army requested donations of high-quality telescopes for observation work in the trenches.

Raids on enemy lines would often be preceded by a reconnaissance by scouts, and they would sometimes be detailed to watch a particular part of enemy lines where a commander thought that enemy activity might be increasing. So effective were the Lovat Scouts Sharpshooters in these roles that a plan was made to allocate groups of them to each division. However, for administrative reasons this did not come fully into force until the end of the war. Whatever the Lovat Scouts Sharpshooters reported was accepted as fact.

Scouting and observing in the First World War had to be adapted to the static nature of the conflict, where opposing forces were often in fixed positions with little movement apart from occasional attacks to take ground. Across the devastated wilderness of No Man's Land, every movement on both sides was noticed by sharp-eyed observers and snipers, and could be fatal. The domination of No Man's Land became part of the psychological battle between the two sides and regular patrols were sent out mostly under cover of darkness both to acquire intelligence and to ambush German patrols. No Man's Land thereby became a place that the enemy became increasingly reluctant to enter. Some of the most effective patrols were carried out by members of the Canadian Corps, which had been formed in September 1915 from the Canadian Expeditionary force, mobilised at the outbreak of war between Britain and Germany.

When the US Expeditionary Force arrived in France in July 1917, it did not bring a cavalry arm with it. Reconnaissance for US forces depended largely on observation balloon companies and about 24 observation aircraft for each division.

Aerial Reconnaissance

Aerial reconnaissance provided the largest proportion of intelligence about enemy movements for high command.

Balloons had been used as early as the Napoleonic Wars for observation purposes and they were also used during the American Civil War (1861–1865) for observation and target acquisition. The Union Army benefited from the advice of one Count Ferdinand von Zeppelin (1838–1917).

Aerial reconnaissance continued apace in the First World War and developments of aircraft were accompanied by developments in camera design.

Inspired by French aerial pioneers such as Louis Blériot, Henri Farman, Louis Boucherou, Gabriel Voisin and Edouard Nieuport, and motivated by their painful experience in the Franco-Prussian War, France forged ahead with developing a military air arm, the *Aéronautique Militaire*, which was established in 1910.

Germany followed in 1916 with the creation of the *Die Fliegertruppen des deutschen Kaiserreiches,* while the British Air Battalion Royal Engineers gave way to the Royal Flying Corps (RFC) in 1912.

By 1912, the *Aéronautique Militaire* was equipped with five squadrons, increasing to 21 squadrons in 1914 and then 65 squadrons. The major focus was reconnaissance, and the Blériot XI was a typical early French reconnaissance aircraft. Another was the Moraine-Saulnier scout monoplane and the Farman MF 11. As the war went on, the number of aircraft dedicated to reconnaissance increased and longer-range squadrons were formed.

Early German aircraft were also almost entirely dedicated to reconnaissance. These included the Aviatik B1 two-seat reconnaissance plane; DFW C.IV, DFW C.V, DFW C.VI, and DFW F37 reconnaissance planes; and the Rumpler-Taube aircraft, whose unusual wing-shape was inspired by the flying seed of the Javan cucumber.

The British Royal Flying Corps began with only four squadrons. Reconnaissance aircraft included the Avro 504, B.E.2 and R.E.8. Photo reconnaissance cameras were fitted to the side of aircraft and from a height of 10,000 feet they might cover an area of about 2 x 3 miles (3.2 x 4.8 kilometres). The British had started an aerial photography unit at the Experimental Flight at Farnborough in 1913, where a Farman aircraft was fitted with a Watson Air Camera that could be vertically mounted on an aircraft and take overlapping photographs. Later developments included the Type A, C, E and L cameras. The work at Farnborough was supplemented by 3 Squadron RFC who devised a way of developing the photographic plates while still airborne, so that they could be handed over immediately to intelligence upon landing.

As aerial reconnaissance and camera work developed, so too did camouflage and concealment techniques on the ground. It became imperative, therefore, to recruit intelligence officers who could interpret the photographic evidence accurately.

Gaining information about enemy dispositions and fortifications was more important than shooting down their aircraft, although the fighter pilots received more attention. For this reason, reconnaissance aircraft themselves were vulnerable to attack, especially when flying in a straight line at a fixed height in order to optimise the photography. They could be escorted by fighters or some were built to fight back. The Bristol F.2B fighter was a case in point. Sturdily built and with a powerful Rolls-Royce Falcon V2 engine, it was armed with a forward-firing .303 Vickers machine gun and a .303 Lewis gun mounted in the observer's cockpit at the back.

When the United States joined the conflict in 1917, they used French aircraft fitted with either French or British cameras, as well as new American camera developments such as the tri-lens camera.

The Inter-War Period (1918–1939)

Amid the seismic changes that followed the war to end all wars, but which unfortunately did not do so, developments in reconnaissance were a mixed bag. Those whose responsibility was to plan reconnaissance activities would not have known at the time that trench warfare on the scale of the Western Front would not be repeated. The role of the aeroplane as a reconnaissance arm was consolidated, but the focus of air forces would move towards strategic bombing and aerial combat. Perhaps the most striking and ominous example of this was the activity of the Italian expeditionary air force, *Aviazione Legionaria*, as well as the German Condor Legion on behalf of the Nationalists under General Franco during the Spanish Civil War (1936–1939). This included not only the famous bombing of Guernica by the Condor Legion on 26 April 1937 but also of Madrid by the *Aviazione Legionaria* in 1936.

The role of cavalry was a matter of discussion. For some, the horse had had its day; others maintained its role alongside increased mechanisation. For some nations the cavalry arm remained responsible for reconnaissance, whereas in others, such as Germany, this hegemony was challenged.

New designs were created for light reconnaissance vehicles, some wheeled and some tracked. In the early to mid-1930s, the Germans developed light armoured cars, such as the Kfz 13 and later the four-ton Sd.Kfz. 221, along with motorcycles, mostly with sidecars. The British also developed scout cars, such as the Daimler Dingo from 1938 and Humber Light Reconnaissance Car and Humber Scout, which entered service in the early 1940s. They also introduced tracked vehicles into their inventory, including the Scout Carrier Mk 1, and Bren Gun Carrier developed since 1934, and light tanks. The Scout Carrier Mk 1 had a different hull layout to the standard Bren Gun carrier. They were fitted with a No. 11 wireless and armed with a Boyes Anti-tank rifle instead of a Bren Gun. However, it would later be replaced by the Universal Carrier.

The British infantry brigade reconnaissance groups would later be consolidated under the Reconnaissance Corps, which was activated by January 1941. Entrance into the corps was demanding, requiring a higher level of IQ than ordinary infantry, and candidates also had to demonstrate initiative and an aptitude for challenges. Those who made the grade regarded themselves as elite. The scout vehicles carried out a similar role to that of cavalry in a previous era. They went forward to screen the main army columns and to spot and report back on enemy movements. Recce squadrons consisted of three scout groups and an assault group.

The French, like the Germans, used both horses and mechanised reconnaissance vehicles. The Soviet Union mechanised their reconnaissance forces, but their systems were still under review when the Second World War began. Their arrangements were not helped by purges of experienced senior officers by Stalin, including the Great Purge which ended in 1939 and ongoing purges of officers through 1940 to 1941, or by his refusal to heed warnings of a German invasion by British, American and Soviet intelligence services.

German Reconnaissance Forces and the War of Movement

Heinz Guderian, a *Jäger* officer, was given responsibility for the development of German mechanised forces in 1922 with no background in the subject. Aware that the British and French had more experience in this area, he acquainted himself with their writings.

> It was principally the books and articles of the Englishmen Fuller, Liddell Hart and Martel that excited my interest and gave me food for thought. These far-sighted soldiers were even then trying to make of the tank something more than just an infantry support weapon. They envisaged it in relationship to the growing motorisation of our age, and thus they became pioneers of a new type of warfare on the larger scale.
>
> I learned from them the concentration of armour, as employed in the battle of Cambrai. Further, it was Liddell Hart who emphasised the use of mechanised forces for long-range strokes, operations against the opposing army's communications, and also produced a type of armoured division combining panzer and panzer-infantry units. Deeply impressed by these ideas, I tried to develop them in a sense practicable for our own army.

Guderian's early explorations involved researching 'the employment of tanks, particularly for reconnaissance duties in connection with cavalry'. In the late 1920s, the Germans translated the English handbook on armoured fighting vehicles, which became the standard text for the German army as they continued to evolve their own ideas. When Guderian was given command of the 3rd

Prussian Motorised Battalion in February 1931, one of his first modifications was the addition of armoured reconnaissance cars and motorcycles, which provided the nucleus of an Armoured Reconnaissance Battalion.

Guderian worked with meagre resources. The Treaty of Versailles in June 1919 had initially allowed them to use only old armoured troop carriers, so they created fake tanks with wood and canvas. They had little or no background on the subject of armoured warfare, other than what they could garner from abroad. However, as early as 1932, Guderian organised a series of exercises to explore the possibilities of the employment of tanks, particularly for reconnaissance duties in connection with cavalry.

As they were now encroaching on a traditional cavalry role, they inquired of cavalry command how they envisaged their future role. The initial reply was that the cavalry would focus on the heavy role, leaving the reconnaissance role to the motorised troops. Panzer reconnaissance battalions were, therefore, assigned this role. A new cavalry commander attempted to regain the lost reconnaissance role but was thwarted. Guderian saw it as a victory of new ideas over the old school represented by the cavalry arm.

Meanwhile, in the United States the 1st Cavalry Regiment (mechanised) was activated under Colonel Daniel Van Voorhies in 1933. This included an armoured troop of 15 armoured cars of the 7th Reconnaissance Squadron.

As they watched developments in Europe, in 1940, the Americans came to the conclusion that they needed an armoured division and, in due course, the 7th Cavalry Brigade was absorbed into the newly formed 7th Armoured Division.

Despite rapidly increased mechanisation, the Americans initially continued to use cavalry in areas that would be inaccessible to vehicles. Hence there were horse-mounted squadrons that could be moved by truck to the area of operations.

On the infantry side, a reformed divisional structure now included a reconnaissance troop.

CHAPTER 6

The Second World War

During the Second World War conventional reconnaissance was carried out with motorcycles, Jeeps, armoured cars and light tanks scouting ahead of large military formations. However, during this period new units were developed, trained and equipped to operate independently sometimes far behind enemy lines for long periods, whether in the deserts of North Africa or the jungles of the Pacific. Special forces were also created and established.

German Reconnaissance

Theory is one thing; practice another. The best laid plans can go awry in battle, as the Allies would discover to their cost, especially in regard to their elaborate defence strategies. From the German perspective, however, having worked hard to achieve change and practise new theories, the new arrangements mostly worked, including the use of reconnaissance. Above all, German reconnaissance in the heat of battle proved to be highly adaptable and imaginative.

During the invasion of Poland in September 1939, the largely mechanised German forces were faced with a Polish army that, although numerically superior, was equipped with out-dated tanks and had a large dependence on horse-mounted cavalry. German reconnaissance forces were kept in reserve to be used in conventional combat when the occasion arose. However, they sometimes fulfilled their role as a spearhead to locate and occasionally engage enemy forces. Thus, Armoured Reconnaissance Battalion 3 advanced to the River Vistula ahead of the 3rd Panzer Division. As both the 3rd and 10th Panzer Divisions continued operations to the north-west of Warsaw, Guderian moved to Bielsk with the advance elements of the reconnaissance battalion in which his son, Kurt, was serving. However, Guderian recognised that the tactics employed in Poland would not necessarily work against the French and British.

During the invasion of the west, German combined arms punched a hole through the *Division Légère Mécaniques* (DLMs), which had expected to hinder what they understood to be the main German advance in Western Europe prior to what they expected would turn into an entrenched infantry battle. The Germans had other ideas and, like a charge of medieval knights, punched a hole through the French line of tanks in May 1940, forcing the whole line to retreat and regroup. French cavalry did not fare any better. The Germans were playing by a different rule book. South-east of Brussels, at Gembloux, German reconnaissance forces consisting largely of armoured cars supported by infantry, evaded French defensive strongpoints and probed forwards looking for ways through.

The French were unaware at the time that the German advance across Belgium was not the main attack, though it was designed to be significant enough to appear to be so. As would later be discovered, the main German assault was an audacious advance through the Ardennes forest in May. To the extent that the French reconnaissance spotted German movements in this area, they would only have seen the advance vehicles in single file and would not have been able to appreciate the mass of armour behind. From the German side, reconnaissance forces did not play a significant role in the Ardennes movement. Once they were out of the woods, however, German reconnaissance units played a significant role in probing forward to detect weaknesses in French defences or to assess the landscape for the best areas of advance for the panzers behind them. This included suitable places to ford rivers. The crossing of the Sauer river in May, for example, was spearheaded by reconnaissance motorcycles and armoured cars. Scout cyclists then crossed the river and assaulted the defenders, forcing them to retreat. This combative attitude by German reconnaissance forces would be repeated throughout the campaign. At La Chappelle the scouts were once again first on the scene and assaulted the French defences rather than waiting for reinforcements. On 12 May, the reconnaissance squadron of the 5th Panzer Division armoured reconnaissance battalion discovered a suitable crossing point on the Meuse at Houx. Having got across, they established a temporary bridgehead while the main battalion crossed the river.

On the French side, advance reconnaissance efforts were not so effective. At Neufchâteau, although French reconnaissance units initially held back the advancing panzers, they were soon bypassed and forced to retreat.

German reconnaissance forces also discovered a crossing point across the Semois river, a tributary of the Meuse. This enabled German armour to cross the river north of Bouillon, placing them north-east of Sedan on 12–15 May.

As the German advance towards the coast continued at pace, the reconnaissance units covered their exposed flanks. However, at Arras on 21 May, they dropped their guard and did not provide warning of a British counterattack. Fortunately for the Germans, the British had not carried out adequate reconnaissance either, and the British thrust failed to achieve its full impact, giving the Germans time to re-group under Rommel's command and hit back. The British and French attack may have temporarily delayed the German advance and encirclement of British forces.

There were other occasions when reconnaissance units dropped the ball. A serious case was at Chalons, where the German reconnaissance unit failed to check the bridge for explosives. The bridge was blown up by the French while German units were crossing.

In the spirit of the German *blitzkrieg*, which was about speed, shock and combined arms, the reconnaissance arms played an essential role as fast, manoeuvrable, adaptable forces that were often effective in combat, while also providing essential information for the forces behind them. Increased contact with the enemy led to an increase in armour and armament. Motorcycles were replaced by half-tracks such as the Sd.Kfz. 250/251 and by light and then medium tanks. New armoured cars included the Sd.Kfz. 234 *Schwerer Panzerspähwagen,* which was later fitted with a turret. The Volkswagen *Kübelwagen,* based on the Beetle chassis, was used as a light soft-skinned scout car.

In the North African (1940–1943) and Russian campaigns (1941–1944), reconnaissance units had wide open spaces in which to range in advance and on the flanks of the main force. They would provide a screen for advancing armoured forces, much as their horse-riding cavalry antecedents would have done. The reconnaissance force was also tasked with providing early warning of an enemy reconnaissance probe or attack. They were highly effective in fending off enemy outflanking manoeuvres when the German forces were in retreat.

In the Russian campaign, reconnaissance forces were the first to cross the Russian border and often probed well ahead of the main force, isolating pockets of Russian forces. In the vast spaces, reconnaissance forces played a key role in identifying the position and intentions of Russian forces, which had a tendency to appear out of nowhere. Apart from the main Russian army, German forces also had to be constantly aware of the threat from partisans who lurked in the deep woods.

French Reconnaissance

French reconnaissance in 1940 may not have been so short-lived if their reconnaissance units had managed to keep a closer watch on the Ardennes region for possible German incursions. As it was, the entire German XIX Panzer Corps managed to get through the forests unobserved. Their engagement by the 2nd and 5th French Light Cavalry proved to be too little too late, and the French were forced to retreat. There would have been little in the training manual about how to delay seven panzer divisions. Although French resistance was brave, they were often overcome by German combined arms, including attacks by Stukas on their artillery positions.

British Reconnaissance

Unable to get back on to the European Continent, the British, and later the Americans, took up the fight against the Axis powers in the form of, first the Italian armed forces, and later the German Afrika Korps in the desert campaign in North Africa (1940–1943).

In the vast open spaces of north Africa, there was plenty of scope for reconnaissance in order to locate the enemy and forestall his movements. The challenges of desert reconnaissance and raids stimulated British minds. Solutions to these challenges often resulted from the inspiration and experience of adventurous individuals, including Ralph Alger Bagnold, OBE, FRS and Lieutenant Colonel Sir Archibald David Stirling, DSO, OBE, who created the Long-Range Desert Group and the Special Air Service respectively.

The formation of the Long-Range Desert Group and the SAS was also influenced by the ideas and exploits of Thomas Edward Lawrence during the Arab Revolt (June 1916–October 1918). While convalescing in 1916, T. E. Lawrence, CB, DSO, who had a background as an archaeologist and knew the terrain of the Middle East well, conceived a new style of warfare that could be used against the Ottoman Turk occupation of the Middle East. He made a virtue of the mobility of the relatively small Arab forces under Emir Faisal and devised 'a thing intangible, invulnerable, without front or back, drifting about like a gas'. The effect of such a force would, like the wind, come or go as it pleased and have the effect of tying down large numbers of enemy troops in nervous expectation.

> Our cards were speed and time, not hitting power … In Arabia range was more than force, space greater than the power of armies.

Lawrence was fortunate to have a commander-in-chief in Field Marshal Allenby, who understood how to harness Lawrence's idea and became one of the early exponents of the use of 'special forces' combined with conventional military strategy. Working with the Bedouin, Lawrence trained himself to match their ways and their ability to survive in extraordinary harsh conditions. Endurance combined with initiative, daring and the ability to win the hearts and minds and harness the abilities of the local tribesmen, made him an inspiration for the modern special forces soldier.

Lawrence and Feisal's Arab forces carried out their raids on camels, which gave them almost unlimited mobility. However, Lawrence also appreciated the value of the Rolls-Royce Silver Ghost armoured cars that were deployed to the region as part of the Machine Gun Corps. Between 1915 and 1918, the British had used armoured cars against incursions by the Senussi, who were allied to the Ottoman Turks. These operations were largely carried out by the Western Frontier Force (WFF) based at Mersa Matruh, along with the Egyptian Expeditionary Force. The WFF included the British Camel Corps, the Australian Light Horse and the Duke of Westminster's Armoured Car Unit. The Egyptian Expeditionary Force included Nos. 11 and 12 Armoured Motor Batteries.

Lawrence swapped a camel for a Rolls Royce, for a raid on a Turkish fort which also involved blowing bridges and tearing up railway lines. They drove away with ammunition and arms captured from the fort.

Light Car Patrols (1916–1919)

The Light Car Patrols were based on Model T Fords, about 40 of which were used to roam the desert on reconnaissance missions and to guard the border from incursions. For this purpose, they were fitted with machine guns. They were also used for mapping uncharted areas of the desert between the river Nile and Siwa. The Light Car Patrols were the inspiration behind Ralph Bagnold's formation of the Long-Range Desert Group in the Second World

War. The patrols included officers such as Claud Williams, who had a talent for surveying.

When Captain Ralph Bagnold, Royal Signals, who had a background as a geologist and explorer, arrived in Egypt in 1920, he mixed with members of the 3rd Armoured Car Company, part of the Royal Tank Corps, several of whom had seen service with the Duke of Westminster's Armoured Car Squadron during the First World War.

In 1929, Bagnold set out on an expedition with a Ford Model A and two Ford lorries to explore the desert between Cairo and Ain Dalla. Apart from his valuable geological discoveries, Bagnold and his companions also discovered much about the use of vehicles in the desert, which would serve him well in the Second World War. They learned how to drive up slopes in high gear and keep the tyre pressures low for better traction in the sand. In the vast empty areas of desert, Bagnold and his team also became experts in navigation, as their lives depended on it. To get round the problem of the magnetic effects on a standard compass caused by the metal of the vehicle, they developed a sun compass that could be used in vehicles to keep them on track. The compass was made from a vertical needle in the middle of a three-inch diameter horizontal disc with 360-degree gradations. Using a card with the sun's azimuths, the disc could be turned every ten minutes or so to provide an accurate reading.

When Bagnold presented his idea of a long-distance reconnaissance force to Archibald Percival Wavell, 1st Earl Wavell, he was asked what else the force could do. The answer was that the LRDG would be equipped to also carry out raids when the opportunity arose, but the quandary among high command about how to make best use of the force remained.

Formed from volunteers, including British, New Zealanders, Rhodesians and Indian servicemen, the LRDG (June 1940–August 1945) was divided into six patrols. Some patrols were organised on regimental lines, for example 'G' Patrol consisted mostly of Coldstream and Scots Guards. Later the LRDG was divided into two squadrons which largely placed the New Zealanders and Rhodesians in A Squadron and the remainder in B Squadron. Later, an Indian Squadron was also formed. An Air section was attached to provide essential back-up and emergency evacuation.

The LRDG travelled mostly in two-wheel drive Chevrolet 1533X2 30cwt vehicles. These were fitted with radiator condensers and typically armed with a Lewis gun and a .303 Browning machine gun. Some vehicles carried Bofors guns or RAF anti-aircraft Vickers machine guns.

Road Watches

The LRDG would position themselves to monitor the movement of enemy reinforcements. One method was to take up a position with the vehicles about two miles from a known Axis route where the vehicles would be camouflaged. LRDG members would then move forward and take up camouflaged positions about 250 metres from the road and take a record of enemy movements. If a large body of enemy or tanks was seen, a radio message would be sent urgently back to headquarters.

In September 1940, 'W' and 'T' Patrols of the LRDG carried out a reconnaissance at Kalifa, following the Italian invasion of Egypt. Having not discovered any signs of movement by Italian forces, they then carried out a raid to the south where they hit Italian supplies. This patrol demonstrated the LRDG's ability to transfer from a reconnaissance to a raiding role, depending on the circumstances.

In December 1940, 'G' and 'T' Patrols of the LRDG liaised with Free French forces to carry out a raid on an Italian fort at Murzuk. In January, they were ambushed by an Italian patrol and lost three vehicles. The LRDG was then largely based at Kufra in Libya, from where it patrolled behind the lines of Axis forces, now including the Afrika Korps. 'T' and 'S' patrols carried out road observations south of the Gulf of Sirte.

From November 1941, under the command of the Eighth Army, the LRDG patrols moved north to central Libya. Here, their role included inserting men of the Special Air Service (SAS) behind enemy lines.

During Operation *Crusader* (18 November–30 December 1941), designed to relieve Tobruk, the LRDG were placed on the offensive to destroy enemy targets. Working with the SAS, the LRDG then moved on to attack enemy aircraft on the ground. This involved driving between rows of parked aircraft and using machine guns and grenades to destroy them.

During Operation *Caravan* in September 1942, the LRDG defeated Italian forces at Barce, in Cyrenaica. This included a successful attack on an Italian airfield, where they destroyed 16 aircraft. However, the LRDG raiding group was ambushed by Italian forces on the return journey, suffering heavy losses.

Special Air Service (SAS)

In circumstances reminiscent of T. E. Lawrence, David Stirling, an officer in the commando section of the British Brigade of Guards, while laid up in hospital, pondered on the advantages and disadvantages of the Long-Range Desert Group and his own experience of parachuting behind enemy lines. One of the problems that Stirling identified regarding the LRDG was that, although it was a highly mobile force, it was also relatively large and liable to be spotted by the enemy. It was also encumbered by relatively large supplies and heavy weaponry. Stirling's vision was for small groups of soldiers to carry out highly targeted raids and then disappear before the enemy could organise retaliation. The idea was to carry out the mission with minimum contact with the enemy, but also to maximise the hitting power of each team member. Stirling envisaged a team of only four men that could have an impact on the enemy of a force ten times its size. The impact would be enhanced by high levels of professionalism.

When it was formed, the new unit was called 'L' Detachment, Special Air Service Brigade, in order to make it seem like it was just another parachute unit. The unit's first mission in November 1941, in support of Operation *Crusader,* was a disaster but it proved its worth on a second mission, this time a ground attack in collaboration with the LRDG. In keeping with the ethos of rapid movement in small teams, the SAS was equipped with Willys Jeeps, which were adapted to carry essential supplies of water and spare fuel, and fitted with twin Vickers machine guns on both the front and rear. The Jeeps were divided into small units, making them difficult to detect by enemy air or ground reconnaissance.

A raid in July 1942 with 18 Jeeps at Sidi Haraish resulted in the destruction of 37 enemy aircraft. The highly mobile force managed to get away and avoid enemy retaliation.

Before Stirling himself was captured by the Germans in January 1943, SAS raids had destroyed about 250 Axis aircraft and a large number of vehicles, while also causing significant damage to railway lines, enemy communications and supply dumps. In Stirling's absence, Paddy Mayne took over command.

While the Long-Range Desert Group had a priority of reconnaissance over raids, the early SAS had a priority of raids over reconnaissance. The role of the SAS in covert reconnaissance will be covered later in this book.

Combined Operations Pilotage Parties (COPP)

This was one of the first military organisations set up for military beach reconnaissance. One of its founders was Lieutenant-Commander Nigel Clogstoun-Wilmott who was a beachmaster during Allied landings at Narvik in northern Norway in April 1940. Later he carried out reconnaissance for a commando raid on the island of Rhodes in the Mediterranean, though the raid itself was called off. Together with Captain Roger Courtney, one of the founders of the Special Boat Service (SBS), they carried out the reconnaissance in canoes wearing special cold-water suits and using infra-red signalling devices. Clogstoun-Willmott also carried out a reconnaissance prior to Operation *Torch*, the Allied invasion of North Africa, which led to the formal establishment of the COPP as a beach reconnaissance unit. During the Allied invasion of Sicily

British soldiers inspecting an abandoned German SdKfZ armoured reconnaissance vehicle in North Africa. (Public Domain)

in July 1943, poor training and equipment led to several casualties but this was soon rectified before the Allied landings in mainland Italy in September 1943.

Operation Overlord *(6 June–30 August 1944)*

The largest seaborne invasion ever mounted saw more than two million Allied troops landing in Normandy between June and August 1944. The preparations for D-Day from August 1943 included a detailed reconnaissance of the coastline. While the Allies encouraged the idea in German minds that an invasion would take place in the Calais area, aerial and surface reconnaissance was carried out on the Normandy coast, aided by intelligence provided by agents from SOE and the OSS and the French Resistance. The Allies were anxious to test the Normandy beaches for their ability to bear heavy vehicles that would be landed in a beach assault, and also to identify the positions of mines, obstacles, gun and machine gun emplacements and beach exits.

Beach reconnaissance was regarded with both scepticism and caution, partly due to the fear of capture and arousing German suspicions about a landing in Normandy. No reconnaissance could be conducted in the Normandy area without the direct authorisation of Winston Churchill himself.

To prove that it could be done, a team of Royal Engineers from the Combined Operations Pilotage Party (COPP) carried out a beach recce in Norfolk, where the beaches were similar to those in Normandy. Swimming from an LCT off the coast, two members of the Royal Engineers successfully evaded sentries, took beach samples and returned to the Landing Craft Tank (LCT) unnoticed.

Having convinced the authorities that the reconnaissance could be safely performed without detection, the Royal Engineers team set out across the Channel in a Motor Torpedo Boat (MTB) on 31 December 1943. Off the coast at Luc-sur-Mer, on a drizzling night, they transferred to a hydrographic craft, while the MTB carried on. At a point beyond the breaking waves, two engineer swimmers, Major Logan Scott-Bowden and Sergeant Bruce Ogden-Smith entered the water. They wore wetsuits and carried equipment such as wrist watches, wrist compasses, torches and a set of 12 tubes each to take beach samples. They also carried waterproof writing tablets to indicate the location from which the beach samples were taken.

A lighthouse was operating nearby, and the men had to lie flat on the beach when the light came near them. It was New Year's Eve and they could hear the sound of German soldiers enjoying the celebrations. Keeping below the high-water mark, they collected their samples before heading back towards the water.

Weighed down by their equipment and the sand samples, they found it extremely difficult to get back out through the breakers, which raised the spectre of capture by the Germans with its implications for D-Day planning. However, they eventually succeeded. The next challenge was rendezvous with the pick-up craft. They flashed their torches out to sea and the craft eventually appeared. In the dark night, and with no lights showing, it took some time for the craft to rendezvous with the returning MTB.

A similar operation was carried out on 18 January 1944 with a COPP midget submarine commanded by Clogstoun-Willmott around the area of Pointe-du-Hoc and the beach that would later be known as Omaha. Through the small submarine periscope, they could see the extensive fortifications and ongoing preparations for gun emplacements, leaving no doubt about the sort of reception that US forces would receive on landing.

At nightfall, two swimmers left the midget submarine and swam ashore to collect beach samples. Despite being spotted by a sentry, they managed to get back to the submarine unscathed. However, the submarine was then surrounded by French trawlers, each of which had a German soldier on board, but somehow the submarine managed to get away undetected.

Operation Market Garden

In September 1944, paratroopers of the 1st Allied Airborne Division landed in Holland to capture key bridges, the most strategically important of which was the bridge over the Rhine at Arnhem. The bridges were to be held pending the arrival of armoured units of XXX Corps. Despite being bravely fought, planning and logistical problems along with a rapid German response meant that Operation Market Garden ultimately failed.

The British 1st Airborne Reconnaissance Squadron played a prominent role in the ill-fated battle for the bridge at Arnhem.

Having seen service in North Africa and Italy, the squadron was no stranger to combat. However, it became embroiled in the mixture of misjudgement and bad luck that bedevilled the Arnhem end of the operation.

The 1st Airborne Reconnaissance Squadron was an elite formation trained and equipped to move fast and send essential information about enemy dispositions and movements back to base. The squadron's movements depended on high levels of co-ordination which in turn depended on good signals communications, both between troops and back to headquarters. Like many reconnaissance forces, the 1st Airborne Reconnaissance Squadron was also regarded as a useful elite light strike unit. The squadron was equipped with both armoured cars and Jeeps but, due to limitations on space in the Horsa gliders for Operation *Market Garden*, only Jeeps were despatched. 45 men, along with Jeeps, were sent in 22 Horsa gliders, while 180 men flew in DC3 Dakotas with the rest of the 1st Airborne Division. This splitting up of the reconnaissance squadron would create the first problem during Operation *Market Garden,* as men and equipment landed in different locations, which delayed their deployment.

General Urquhart's plan was that the 1st Airborne Reconnaissance Squadron should carry out a fast assault on the Arnhem road bridge and await reinforcements. However, Urquhart received a message that most of the reconnaissance Jeeps had not landed and he tried to get in touch with the commander of the 1st Airborne Reconnaissance Squadron, Major Freddie Gough, to find out what was going on. He could not get through to Gough as the recce squadron operated on a different radio net. Although Gough tried to get through to Urquhart by radio, by that time, Urquhart was in his Jeep and unobtainable. Gough then decided to go and find Urquhart and talk to him in person, leaving his unit leaderless. Meanwhile, the recce squadron set off for its objective but soon ran into an SS Panzer Grenadier Training Squadron and Reserve Battalion, under SS Major Josef 'Sepp' Krafft. The lead Jeeps were destroyed while the remainder returned to their start point. Despite the fact that there were two troops available, no attempt was made to find a way around the Germans and scout an alternative route to the bridge.

When Urquhart's personal Jeep was destroyed and his radio operator injured, both the reconnaissance squadron and the entire 1st Airborne Division were cut off from their leaders.

Although much emphasis has been placed on problems with radio crystals in the British number 19 and 22 radios during Operation *Market Garden*, this appears to cast a veil over deeper problems of planning and leadership. The limitations of the radio sets were understood when the operation was planned, especially in view of the wide area to be covered and the distance of the drop zones from the objective. In view of the potential communications difficulties, better fall-back planning such as the use of runners rather than relying exclusively on radios may have mitigated the misunderstandings that occurred on the ground. In any event, the fact that Urquhart and Gough decided to leave their respective commands in order to see for themselves, rather than sending messengers such as despatch riders, added to the growing confusion. As the German commander Field Marshal Model recognised, paratroopers were at their most vulnerable in the period immediately after a jump when they need time to find their units and equipment. In such circumstances, high levels of initiative and local command are required. Although it might seem eccentric, the use of hunting horns and bugles by Colonel John Frost and his deputy Major Allison Digby Tatham-Warter, DSO, to rally their men and send messages turned out to be highly appropriate and effective.

'Old Men on Bicycles'

The 9th SS Panzer Division 'Hohenstaufen', commanded by *Obersturmführer* Walter Herzer, was lucky enough to escape from the battle of the Falaise Pocket after the D-Day landings in Normandy, having previously inflicted heavy losses on British armour. Exhausted after fighting rear-guard actions during their retreat through France and Belgium, the division was sent to Arnhem for repair and refitting, where it came under the command of General Bittrich. When the British Airborne Division landed during Operation *Market Garden*, only the reconnaissance battalion of the *Hohenstaufen* Division was available for action. Commanded by *SS-Hauptsturmführer* Viktor-Eberhard Gräbner, the unit consisted of 40 mostly wheeled or half-tracked vehicles, including the latest Sd.Kfz. 234 Puma armoured car.

Gräbner received an order from Bittrich on 17 September to hold the Arnhem bridge and carry out a reconnaissance of the Nijmegen bridge. However, Gräbner appears to have misunderstood the order, leaving only a token force at Arnhem and taking his main force to Nijmegen, which was already in German hands. A report was sent to Model's headquarters informing him that the road between Nijmegen and Arnhem was clear of the enemy. When Bittrich was informed by other sources that the northern end of Arnhem bridge was held by British paratroopers, he personally went to Herzer's command post to reprimand him for Gräbner's error.

Gräbner was given an opportunity to vindicate himself by evicting the British paratroopers and opening Arnhem bridge so that it could be used to send reinforcements where they were needed. At Gräbner's signal, the armoured cars and half-tracks of the reconnaissance battalion accelerated across the bridge at 9.00am on 18 September. The British paratroopers held their fire until the German vehicles were close by, and then unleashed a hail of mortar, machine-gun and rifle fire. Some of the vehicles collided and caught fire and at least one crashed through a barrier and fell off the bridge. Trapped in the chaos of burning vehicles and exposed to fire from above, the German troops had little chance and the attack failed.

General Principles of US Military Reconnaissance in the Second World War

To the dismay of the cavalry die-hards, US cavalry reconnaissance became entirely mechanised from the summer of 1942. Initially, this

---

US Army Truck, ¼-ton, 4x4, Command Reconnaissance (Jeep)

Light, manoeuvrable and relatively fast, the Jeep was in many ways an ideal reconnaissance vehicle. Although it was unarmoured, its defence was clear observation (it was rarely used with its canvas top on scouting duties) and speed. The Jeep replaced the motorcycle as the standard light vehicle for reconnaissance units and it could be fitted with a wide variety of weapons, including the M19191A4 Browning machine gun, twin Vickers machine guns or 60mm mortar. Scout Jeeps often carried extra cans of fuel in panniers fixed to the outside of the vehicle.

---

meant replacing horses with Harley Davidson motorcycles, but these were found to be too limited for cross-country manoeuvres and were replaced by the quarter-ton truck, commonly known as the Jeep. The Jeep was fully fielded by 1942. Initially, the Jeeps were supplemented with White Scout Cars and half-tracks but, as it became clear that greater protection was required for reconnaissance missions, M8 armoured cars were brought in from 1943. The relatively light armour and armament of the mechanised cavalry reconnaissance units underlined their primary mission, at least in theory, which was reconnaissance and not engagement with the enemy.

US Army doctrine specified, in 1943, that the organisation, equipment and training of mechanised cavalry was reconnaissance, fire-and-manoeuvre and infiltration tactics. Combat was to be minimal and only to the extent that the primary mission required. The cavalry group was expected to provide reconnaissance for the routes that the main force would use, and this would be achieved either by road or across country. Typically, they would report on the state of roads and bridges, potential ambush areas, alternative crossing points where bridges were either destroyed or thought to be booby trapped, and possible safe encampment sites. During an attack by the parent unit, the reconnaissance force might be deployed to explore flanking opportunities.

If the enemy were to retreat, the reconnaissance force might maintain contact and report its movements and any new defensive positions that it might take up.

If the enemy were approaching, the reconnaissance unit would be responsible for providing early warning and for preventing the enemy reconnaissance unit from obtaining information about the friendly force.

Although the official doctrine declared otherwise, reconnaissance units often found themselves engaged in combat. As the first on the scene, they were the most likely to run into ambushes and would need to be combative in order to extract themselves. However, getting bogged down in a firefight meant that they were compromised from fulfilling their primary mission. They were also almost invariably inadequately armed. The standard issue carbine, for example, had neither the range nor the hitting power to make a lasting impression on the enemy.

Ideally, reconnaissance units would always avoid contact with the enemy and would avoid being seen themselves. This would give them the maximum opportunity to report on what was going on both before and after contact with the enemy by the main force. If possible, the reconnaissance unit would identify the location of enemy positions, their weaponry, the size of the force, including the number of tanks and supporting vehicles, danger areas such as minefields and any signs of their planned movements.

On occasion, reconnaissance units would supplement their information by reconnaissance by fire. This would provoke the enemy to respond, on the assumption that it was under attack, and thus reveal the positions of its men, machine-gun emplacements and artillery positions. This information could give advance warning to the advancing friendly force units, so that they were less likely to be surprised by unexpected fire. It could also provide them with target opportunities for their mortars, machine guns or artillery.

Each member of a reconnaissance unit had to be able to function independently, as well as working successfully with the team.

Although the principles of special forces teams were still embryonic, reconnaissance units had to be able to multitask beyond their particular roles, so that each team member could take on any tasks, whether it was driving one of a range of vehicles, working radios or manning weapons.

Battle of Lanzerath Ridge (16 December 1944)

In the opening hours of the campaign in the Ardennes (December 1944–January 1945), often described as the battle of the Bulge, a single US reconnaissance unit managed to halt the German advance for about 20 hours, causing significant disruption to German plans and, perhaps, altering the whole course of the battle in favour of the Americans. The episode is also an example of how reconnaissance units were often caught up in combat incidents for which they were not intended or trained, but due to their ingenuity and pluck came out fighting.

The 394th Intelligence and Reconnaissance platoon of the 99th Infantry Division was deployed to patrol a gap in the US defences south of Losheimergraben. The southern part of the area was patrolled by the 18th Cavalry Squadron, part of the 14th Cavalry Group. The 394th I&R platoon consisted of 18 men and was commanded by 20-year-old Lieutenant Lyle Joseph Bouck, Jr. The unit was also accompanied by four artillery observers. Although few in numbers and relatively inexperienced in battle, the members of the 394th I&R platoon were excellent marksmen, well trained and in optimum physical condition.

The platoon fulfilled their reconnaissance and intelligence-gathering role, for which they were trained. They carried out regular patrols in their area and sometimes went behind enemy lines to capture German soldiers for intelligence. Through their listening and observation posts, they kept headquarters informed of any enemy movements. They carried out covert patrols to fix enemy positions.

On 10 December, the recce platoon was ordered to move to Lanzerath ridge in order to fill the five-mile gap between the 100th and 99th Divisions. They were supported by 22 men of the 820th 2nd Reconnaissance Platoon. They reinforced existing foxholes just behind a tree line, giving themselves standing firing positions and good cover. An armoured Jeep with a .50 calibre machine gun was dug in to defend the likely route for a German advance. Concealed by fresh snow, the position was invisible. Digging in and defending a fixed position was not what the recce platoon was trained to do, but they made a very good job of it nonetheless.

On 16 December, the platoon heard the distinctive clanking sound of approaching German armour. They did not know it at the time, but they were directly in the way of the 1st Panzer Division and the northern thrust of the German advance through the Ardennes towards Antwerp, which would develop into the battle of the Bulge.

Due to heavy losses after D-Day, the German troops of the 3rd Battalion, 9th *Fallschirmjäger* Regiment accompanying the panzers were relatively inexperienced. This soon became apparent as they attempted a frontal attack on the American position, across a field bisected with a barbed wire fence. As they attempted to negotiate the fence, they were cut down by withering fire from the American position. Eventually, Joachim Peiper, the commander of Kampfgruppe Peiper, which was equipped with the latest King Tiger tank, arrived on the scene and demanded to know the reason for the delay. He was told that the entire forest was crawling with Americans.

Although Bouck asked HQ if his reconnaissance unit could make a fighting retreat, he was told to stay in position and await reinforcements. These were not forthcoming and, eventually, the Germans found a way to flank the American position. The Amercans who were not killed were captured and sent on a hellish train journey for days without food or water to a concentration camp. Several died on the way.

The 394th I&R Platoon had stalled the advance of Germany's most powerful tanks, delaying the attack by Kampfgruppe Peiper by about 18 hours. Only after the war ended would the surviving members of the platoon realise what they had achieved, and they became one of the most decorated units by size in the US Army.

US 6th Army Special Reconnaissance Unit, 'Alamo Scouts', Pacific Theatre

During the operations of the US 6th Army in the south-west Pacific from January 1943, its commanding officer, Lieutenant-General Walter Krueger, recognised the need for a specialist reconnaissance force that would keep the 6th Army abreast of Japanese movements in New Guinea and New Britain. Established in November 1943, the new special reconnaissance unit consisted of specifically selected men who were given intensive training. Due to Krueger's enthusiasm for the famous defence of the Alamo in 1836 during the Mexican wars, they soon became known as the Alamo Scouts.

A team from the US 6th Army Special Reconnaissance Unit, or Alamo Scouts. (Wikimedia Commons)

> Formed under the Office of Strategic Services (OSS), Detachment 101 was a special unit that was dropped behind enemy lines in Burma to gather intelligence and to harass the enemy. It also identified targets for Allied bombers and helped to rescue Allied prisoners of war. The unit successfully enlisted the help of the Kachin people in operations against the Japanese.

The Alamo Scouts Training Center pioneered many of the selection processes that are now standard in special forces selection and training establishments. From the hundreds of keen recruits who applied to join the new unit, only 138 graduated as Alamo Scouts.

The training programme included amphibious reconnaissance, patrolling, clandestine and jungle warfare, the use of a variety of craft from rubber boats to torpedo boats and submarines, as well as Catalina flying boats for infiltration into enemy territory.

Although mainly a reconnaissance unit, the Alamo Scouts were also trained and equipped for direct action and for operations such as hostage rescue.

The Alamo Scouts worked with the 6th Ranger Battalion to carry out a raid on a prisoner-of-war camp at Cabanatuan on the Philippines, in January 1945. In this operation, the Alamo Scouts carried out reconnaissance for the Rangers before they went in. Part of the reconnaissance included setting up an observation post in a shack near a Japanese guard post from where the scouts monitored Japanese movements. The raid was successful and about 500 Allied prisoners of war (POWs) were liberated.

The Alamo Scouts conducted over 100 missions in the Second World War, and the unit was disbanded in November 1945. However, several members of the Alamo Scouts joined the US Army special forces.

1st Special Service Force 'Devil's Brigade'

Formed in July 1942, this special force unit included both American and Canadian servicemen. After a couple of aborted missions, it became heavily engaged during the battle of Monte Camino, on 5 November 1943, with a special mission to take Monte La Difensa. This involved an intrepid climb up a cliff and an assault on the German position. They were also deployed at Anzio, where they conducted aggressive reconnaissance patrols deep behind enemy lines, causing the Germans to rush reinforcements to the area. Their covert operations at night, with blackened faces, earned them the nickname from the Germans as the 'Devil's Brigade'. The unit also took part in Operation *Dragoon*, the Allied invasion of southern France on 15 August 1944, and they were later attached to the 1st Airborne Task Force.

US Marine Corps Scout and Sniper Unit

In September 1942, while based on Guadalcanal, the US 1st Marine Division formed a scout and sniper detachment to carry out reconnaissance and similar duties, such as artillery observation.

The selection process focused on those Marines with skills in fieldcraft and marksmanship. Although they initially conducted patrols on foot, the unit was later reinforced with tanks for reconnaissance in force operations.

The development of the scout sniper concept would prove to be enduring in the Marines and distinguished them from military units where the sniper's role was focused on acquiring and eliminating targets.

United States Marine Corps Amphibious Reconnaissance Battalion

The US Marine Corps Amphibious Reconnaissance Battalion played a key role in the island-hopping campaign by US and Allied forces in the Pacific between 1943 and 1945. They often worked with the US Navy Underwater Demolition Teams (UDTs).

The reconnaissance battalion was trained in infiltration by several different methods, including a variety of watercraft and torpedo boats, high-speed destroyer transports or by Catalina flying boats.

M8 Greyhound light armoured cars of US 25 Cavalry Reconnaissance Squadron passing through Foligny in Normandy on 31 July 1944. (Wikimedia Commons)

The main purpose was to perform accurate surveys of beaches to test their viability for landing heavy military equipment. This would include the load-bearing capacity of the beach, natural or man-made obstacles and other relevant topographic and hydrographic information. Their reconnaissance and scouting duties also included the location of enemy defences and routes for advance inland.

Operation Jedburgh *(June–December 1944)*

A cooperation between the British Special Operations Executive (SOE) and the American Office of Strategic Services (OSS), the *Jedburgh* operations began just before and after the Allied landings in Normandy, and were designed to coordinate the activities of local resistance movements with the advancing Allied forces. The teams were made up of three military personnel, one of whom would be either British or American, another of the nationality of the country of operations and a third who could be of either nationality. The teams were uniformed military personnel and the operations by that stage of the war were designed to be overt rather than covert. Operations often involved sabotage, as well as providing information about enemy activity.

The *Jedburgh* teams were an important stepping stone in the formation of special forces. Colonel Aaron Bank, who was parachuted into France with one of the teams, would later help to found the US Army Special Forces, often known as the 'Green Berets'.

Australian Service Reconnaissance Department (SRD)

The Australian Service Reconnaissance Department developed from the Inter-Allied Services Department (ISD), formed in April 1942. The SRD in turn was under the control of the Allied Intelligence Bureau (AIB). In due course, the SRD was replaced by Special Operations Australia, which acted independently. The SOA was responsible for reconnaissance and intelligence-gathering operations in Japanese-occupied countries.

Australian Z Special Unit

Inspired by the British Special Operations Executive (SOE), the Australian military formed a Reconnaissance Department in March 1942 to oversee intelligence gathering and raids behind enemy lines.

The Z Squadron unit was the active reconnaissance and special forces unit of the SRD and it was trained in both parachute and submarine infiltration to carry out reconnaissance and carry out raids.

During Operation *Sedgwick* in September 1943, the Z Special Unit attacked Japanese shipping in Singapore harbour, sinking six ships. The 11 Australian and three British operators moved into the region in a vessel disguised as a local fishing boat. They then paddled in collapsible canoes for about 50 kilometres to a rendezvous on the coast, where they waited until nightfall for the raid on the harbour. Having entered the harbour, they placed limpet mines on several ships before returning to their rendezvous and making their way back to Australia. As a result of the raid, seven Japanese ships were either damaged or sunk.

M Special Unit

The M Special Unit was formed in 1943 to provide reconnaissance and carry out raids in the south-west Pacific during operations against Japan. Formed from personnel from the Australian, New Zealand, British and Dutch forces, the M Special Unit operated in small teams, which were inserted covertly behind enemy lines to provide key information on Japanese movements.

CHAPTER 7

The Cold War Years, 1950–1982

During the period of the Cold War, there was a marked development in military reconnaissance especially in the wars and conflicts in south-east Asia where reconnaissance techniques and units had to be adapted to cope with challenging environmental conditions and highly elusive foes.

The Korean War

When the Democratic People's Republic of Korea (North Korea) invaded the Republic of Korea (South Korea) in June 1950, the United Nations became involved in the conflict, with the United States contributing with the largest forces.

North Korea's Korean People's Army (KPA) was a formidable force modelled on Soviet mechanised military lines and equipped with Soviet armaments. By contrast, the South Korean Army was relatively weak and unprepared. The US Eighth Army in the region was also lacking in equipment and manpower after 1945.

United States Marine Corps 1st Reconnaissance Battalion (1st Recon Bn)

The US Marine Corps, along with the US Navy in general, had suffered drastic cuts after the end of the Second World War, and was relatively unprepared for another major conflict. However, a US Marine division was ordered to Korea as part of the UN command, stimulating a race to rebuild Marine strength. On 3 August 1950, the 1st Provisional Marine Brigade landed at Busan in South Korea, coming under the overall command of the US Eighth Army under Lieutenant-General Walton Walker. The rest of the US Marine Corps Division landed at Incheon.

Although primarily trained for amphibious infiltration, the 1st Recon Bn were tasked with Jeep-mounted reconnaissance patrols into enemy territory, primarily covering ports on the east coast. Other patrols included identification of enemy infiltration as well as infantry deployments and the direction of air strikes.

The 1st Recon Bn was also employed in its more familiar role of amphibious infiltration from the high-speed transport ship USS *Horace A. Bass*. They were accompanied on some raids by US Navy Underwater Demolition Teams (UDTs), attacking North Korean infrastructures, such as tunnel and railway bridges.

Malayan Scouts (SAS)

The Special Air Service continued to exist as a territorial unit after the Second World War and it was the communist insurgency in Malaya beginning in 1948 that would lead to its rebirth as a fully-fledged special forces unit for the post-Second World War world.

The Federation of Malaya had been formed from several former British colonies and continued to be ruled by a colonial government, which respected the rights of local rulers. The mostly Chinese communist party of Malaya wanted to impose independence under communist rule and promoted an insurgency to achieve this.

From 18 June 1948, a state of emergency was declared, and the British and Commonwealth authorities attempted to counter the insurgent threat. Although initial efforts using standard military practice were unsuccessful, gradually, a counter-insurgency methodology was developed, in which the Malayan Scouts (SAS) played a central role. Rather than trying to crush the insurgency with overwhelming force, the Malayan Scouts worked with local communities, learning their language and caring for their needs, including medical. Communities were protected from the insurgents. Alongside this, long-range reconnaissance patrols by the Malayan Scouts put the insurgents on the back foot and made it clear that they could not control the region. During this period, the SAS developed patrol methods that would stand it in good stead for the future.

Borneo Confrontation – Operation *Claret*

The SAS did not have long to wait for another opportunity to practise their long-range reconnaissance skills. After the creation of the Federal State of Malaysia, on 31 August 1957, ethnic Chinese supported by Indonesian forces carried out incursions on the island of Borneo from Indonesian held Kalimantan into Sarawak.

Although hostilities were not officially declared, under Operation *Claret* the British Special Air Service (SAS) and Special Boat Service (SBS), Australian Special Air Service Regiment (SASR) and New

A patrol of the Australian Special Air Service Regiment in a Land Rover Perentie. (Australian Defence Force)

Zealand Special Air Service (NZSAS), were supported by elite and conventional forces including the Royal Marines, the Parachute Regiment, Royal Australian Regiment, Royal New Zealand Infantry Regiment and the Gurkhas.

Long-range reconnaissance patrols by the special forces were designed to penetrate enemy territory and identify the size and movements of enemy units. Psychologically, it also sent a message to enemy insurgents and military patrols that they could not move in safety. During this period, the SAS refined its four-man patrol structure, which allowed them to move covertly to carry out their reconnaissance missions. They also employed the help of local people, such as the Dayaks, who had excellent knowledge of the region and who helped them to improve their jungle skills.

A change in the Indonesian regime in 1965 meant that the Indonesians no longer supported the insurgency, which helped Commonwealth forces to bring the insurgency to a close.

The Vietnam War

The Vietnam War, which lasted from about November 1955 to April 1975, was similar to the Korean War to the extent that a communist North wanted to unify the country by invading a non-communist South backed by the United States and its allies. Apart from North Vietnam, the South Vietnamese government also faced an insurgency by the communist Viet Cong in South Vietnam.

US military participation in the conflict began to increase considerably in 1961, with combat units deployed in 1965. By the end of the decade, there were over 500,000 United States military personnel in Vietnam. Support by the Soviet Union and China for North Vietnam and the communist insurgency also increased exponentially. Despite its massive military investment, however, by the early 1970s the United States recognised that victory would continue to be elusive and it withdrew its forces in 1973. South Vietnam fell to the North Vietnamese forces in 1975.

One of the major challenges that US forces faced in Vietnam was fighting an elusive and escalating insurgency in dense jungle terrain. This insurgency by the Viet Cong (VC) in South Vietnam was backed and supplied by the North Vietnamese Army (NVA). The US Joint Chiefs of Staff recognised that a sophisticated counterinsurgency and reconnaissance force was required in these circumstances to detect and, where possible, counter enemy movements through target acquisition and forward air control (FAC).

The insurgency was sustained by a constant supply of armaments, ammunition and other supplies which were delivered via several trails from the north, the most notorious being the Ho Chi Minh trail. An initial trickle of supplies would eventually escalate to the movement of entire NVA divisions into South Vietnam.

The Ho Chi Minh trail, though named in the singular, was in fact a network of trails, which meant that if one part of the trail was bombed an alternative route could often be used. Since it was a trail and not a road, it was also relatively easy to repair by filling in bomb craters. The trail was often covered by a dense jungle canopy and supply dumps were camouflaged, making them difficult to spot from the air. In view of all this, ground reconnaissance was the only truly effective method of identifying movements on the trail, as well as the location of supply dumps and anti-aircraft installations.

Military Assistance Command, Vietnam-Studies and Observation Group (MACV-SOG) (January 1964–May 1972)

MACV-SOG was a highly classified product of joint services command with a range of duties, including reconnaissance, psychological operations, hostage and downed airmen rescue and direct action. It included personnel from several elite and special operations units from the navy, army, marines and air force, as well as the Central Intelligence Agency (CIA).

MACV-SOG teams would typically operate out of a Forward Operating Base (FOB) and would be inserted into their location over the border by either US Marine Corps or South Vietnamese aircraft. Their task was to provide intelligence for as long as they could stay in the location, taking into account the risks involved, including discovery by NVA counter-reconnaissance operations. Radio communications would be made with a Forward Operating Aircraft (FAC). This aircraft would pass on target locations to strike aircraft.

MACV-SOG's operations extended into Laos and Cambodia and, in due course, their operations were affected by the increase in conventional manoeuvres by the NVA, including the Tet Offensive in 1968.

The unit was stood down in January 1973, and its existence remained clouded in secrecy for decades to come. However, its operations added a great deal to the knowledge base and experience for the ongoing development of US special forces.

Long-Range Reconnaissance Patrols (LRRPs) in Vietnam

US forces had gained considerable experience of long-range patrols behind enemy lines through the Alamo Scouts in the South-West Pacific during the Second World War, which was consolidated by the Alamo Scouts Training Center.

The concept of long-range reconnaissance operations was resurrected in the US forces in the 1960s, partly as a result of NATO doctrine, which in turn was influenced by British special operations experience, although long-range reconnaissance, with its emphasis on intelligence gathering and target acquisition, was distinct from special operations. British long-range reconnaissance forces tended to emphasise static observation, while American practise tended to be more mobile.

During the Vietnam conflict, the United States came to realise that neither the Viet Cong (VC) nor the North Vietnamese Army (NVA) had read the script so far as conventional warfare concepts were concerned, such as gaining ground and holding it, and therefore growing emphasis was placed on long-range patrols to ascertain the position and movements of the elusive enemy.

The designations of long-range reconnaissance patrols changed rather confusingly over the course of the Vietnam War, and the way they were employed also tended to change, depending on who happened to be in charge at the time. Broadly, they were called LRRP, LRP or Ranger patrols. Their mission was primarily reconnaissance, though their direct combat actions against the enemy tended to increase over time.

A typical task list for an LRRP of 1st Brigade, 101st Airborne might include some of the following: determine the location of the enemy in a particular area; report on their supply routes; provide information on which enemy units were operating; capture the enemy for interrogation; and establish a forward observation post for fire support and set up an ambush. Although the primary aim was gathering intelligence, taking prisoners and setting up ambushes, they sometimes had direct contact with the enemy.

Depending on operational requirements, an LRRP unit might be tasked with more specific missions focused on enemy assets or installations, such as finding and destroying an enemy radio transmitter or locating the position of an enemy anti-aircraft gun. Weapons caches and hospitals were also on the bucket list.

LRRP units were partly victims of their own success. As commanders at HQ appreciated their intrepid and mostly successful missions, they also became inclined to use them in special operations roles to hit the enemy rather than watch him. The commander of the 25th Infantry Division was explicit about changing the emphasis of the LRP and Ranger divisions to an offensive role. He would include ambush and rescue missions. The teams were given heavier armaments, including M60 machine guns, to reflect this change of emphasis, and a sniper was included as part of the team. However, this successor reversed the emphasis back to reconnaissance.

The 4th Infantry Division sometimes used LRPs to secure a landing area for a combat mission. The LRP patrol would be inserted up to five kilometres from the landing zone and then walk in towards it. This gave the combat teams greater confidence that they were not

A reconnaissance troop of UK 4th Infantry Brigade mounted in Jackal-armoured wheeled vehicles. (Ministry of Defence)

going to get jumped after landing, and it also obviated the need to spray the surrounding jungle with precious ammunition.

Depending on the location of the divisional LRP units, terrain and population factors also played a significant role. It was one thing to be inserted covertly into a densely forested area; quite another to enter an area with little cover and with a significant local population, any of whom could harbour enemy spies or pass on information.

Sometimes LRP units would work alongside special operations units, such as US Navy SEALs. They would also work with South Vietnamese forces, paratroopers and sniper teams.

Australian Army Reconnaissance Platoon

Australia sent a small team of advisers to Vietnam in 1962, and, by the end of the decade, the Australian military involvement in the region had increased to over 7,000 personnel.

In 1966, it was recognised that, due to the particular nature of the terrain and the enemy, reconnaissance should be increased ahead of advancing regular forces. Like many reconnaissance teams, the purpose of this platoon would be to gather tactical intelligence and report back to headquarters, only engaging with the enemy if necessary.

The new unit received training from 3 Squadron Special Air Service.

Spetsnaz

The Russian Spetsnaz, whose name indicates 'special purpose', were developed to counter a perceived threat from US tactical nuclear weapons in Western Europe. The concept of special reconnaissance was developed to locate such weaponry and to neutralise it, along with identifying and destroying any other perceived threats. However, Spetsnaz are also trained in conventional tactical reconnaissance, as well as target observation.

Another term for the work of the Spetsnaz is *Diversiya rezvedka* or 'diversionary reconnaissance', which covers the identification of assets vital to NATO movements, such as bridges and supply depots, and destroying them.

A Spetsnaz unit typically operates a team of eight, among which one would be the reconnaissance expert, and others the lead on sniping, explosives and communications. However, like all special forces teams, each would also be trained in each other's area of expertise.

Arab–Israeli Wars

During the Sinai Campaign (Operation *Kadesh*) of 1956, the Israeli Defence Force deployed a reconnaissance force mounted in Jeeps, sometimes receiving support from AMX-13 light tanks. The Jeeps were armed with anti-tank recoilless rifles as well as machine guns.

At the strategically important pass of Abu-Ageila, reconnaissance by Jeep-mounted companies resulted in the identification of a relatively undefended pass which allowed the Israeli forces to pass through and gain an advantage over the Egyptian defenders.

By 1967, although Jeeps continued to be used, half-track armoured personnel carriers were added to the reconnaissance company inventory. These were armed with .50 calibre machine guns, 20mm cannons and anti-tank guns.

During the Six-Day War of 5–10 June 1967, aggravated by continuing tensions between Israel and its Arab neighbours, particularly over the closure of the Straits of Tiran to Israeli shipping, Captain Ori Orr led the Reconnaissance Company of the 7th Brigade in a significant action at Rafeh Junction where, despite being outgunned by Egyptian forces, Orr's company managed to fight back and cause the Egyptians to retreat. The Reconnaissance Company continued to lead the IDF advance as it headed west further into Egyptian territory. However, the Jeeps of the reconnaissance company were badly mauled by Egyptian artillery in another ambush near Jiradi on 6 June 1967, highlighting the vulnerability of unarmoured light scout vehicles. By 9 June, IDF reconnaissance forces had reached the Suez Canal, acting as a spearhead for the Israeli advance.

1973 Yom Kippur War

By the time of the next Arab–Israeli conflict, and in recognition of their vulnerability in previous battles, Israel had upgraded its reconnaissance forces with M113 armoured personnel carriers and tanks, although a Jeep-mounted scout company still played a role in the battalion organisation.

Reconnaissance units played a key role in discovering gaps in the Egyptian defences which enabled Israeli forces to cross the Suez Canal undetected. However, the reconnaissance unit came under heavy fire, once again raising the question of how heavily armoured reconnaissance units should be.

Israeli General Staff Reconnaissance Unit 269 – Sayeret Matkal

This special operations unit was established from 1957 to gather strategic intelligence behind enemy lines. It developed in due course into a hostage-rescue and counter-terrorist force, making headlines with the Entebbe hostage rescue raid, in July 1976.

Falklands War, 2 April–4 June 1982

Argentina, which had disputed British sovereignty over the Falkland Islands since 1833, mounted an invasion of the islands on 2 April 1982. The British operation to take back control of the Falkland Islands, Operation *Corporate*, was in the face of daunting political, military, logistical, geographic and meteorological challenges. It might be said that the British were fighting two enemies – the Argentine armed forces and the Falklands weather. The cash-starved and ill-quipped British military had to send a task force almost 8,000 miles into the South Atlantic, where they would meet the Falkland Current, known for its Antarctic-like low temperatures. The islands themselves were hilly, with peat and marsh valleys, and almost constantly windy, with a mean annual temperature of 5°C. There was not much chance in the Falklands of drying your wet socks in the sunshine. The Falkland Islands consisted broadly of east and west Falkland, each island being cut with numerous inlets.

However, the cobbled-together force, including an aircraft carrier snatched back from sale to India and a civilian supply ship, included some of the world's most professional military personnel. Ground forces included three Royal Marine Commando battalions, two Parachute Regiment battalions and both the Special Boat Service and 22nd Special Air Service Regiment. Infantry included both the 2nd Battalion Scots Guards, 1st Battalion Welsh Guards and Gurkha Rifles.

A Royal Marines officer in the task force, Lieutenant-Colonel Ewen Southby-Tailyour, OBE, had previously sailed solo round the Falkland Islands' coast and taken extensive notes. These would prove invaluable for the British in landing special forces teams and major landings of troops and supplies.

Among the special forces, the Special Boat Service (SBS) were tasked with shore reconnaissance, and the Special Air Service (SAS) with inland reconnaissance. Reconnaissance for the Royal Marines was largely carried out by the Arctic Warfare Cadre.

On 1 May 1982, four-man teams of G-Squadron SAS were inserted on the Islands to observe Argentine positions and to call in airstrikes. Due to weather conditions and navigational difficulties, the teams were dropped over four miles from their objectives. Once on the ground, they were faced with the difficulty of concealing themselves in the open rocky terrain with almost no cover. They dug scrapes and concealed themselves and their equipment with camouflage netting.

Operation Plum Duff

One of the greatest threats to the British task force was the French-designed Exocet missile, delivered by French-designed Super Étendard fighter bombers. British ships were being sunk at an alarming rate. These included the Type 42 destroyer HMS *Sheffield*, hit by an Exocet missile on 4 May; the Type 21 frigate HMS *Ardent*, sunk by bombs on 21 May; the Type 21 frigate HMS *Antelope* sunk by bombs on 23 May; the Type 42 destroyer HMS *Coventry*, sunk by bombs on 25 May; the converted container ship SS *Atlantic Conveyor*, sunk by Exocet missiles on 25 May; and the Royal Fleet Auxiliary ship *Sir Galahad*, sunk by bombs on 8 June. British commanders' attempts to neturalise the aerial threat included a plan to fly an SAS team to the Argentine mainland air base at Tierra del Fuego to destroy the Exocet missiles and the planes that carried them. Called Operation *Mikado*, the plan would be preceded by a reconnaissance operation, called *Plum Duff*.

An eight-man team from B Squadron SAS flew in a Sea King of 846 Naval Air Squadron on 18 May. The aircraft soon ran into thick fog, causing navigational problems, which in turn raised the issue of the Sea King's limited flight range. The helicopter landed across the border in Chile, where the crew attempted to destroy it. Meanwhile the SAS team planned to head over the Argentine border to complete the recce operation. However, the helicopter was spotted, and the mission was called off now that its cover was broken.

Mount Kent Reconnaissance (25 May 1982)

Having been advised by patrols from G Squadron SAS that some of the high ground around Port Stanley in the Falklands was lightly defended, a more thorough reconnaissance of the area was carried out by D Squadron SAS. However, the Argentine Army had a similar idea and the SAS found themselves engaged in firefights with the Argentine commandos. The SAS managed to maintain a toehold on Mount Kent until they were reinforced by Royal Marines.

The SAS reconnaissance mission had been important because, otherwise, the Royal Marines Commandos would have landed in the midst of the Argentine commandos without warning, leading to probable losses of both aircraft and men.

Royal Marines Arctic Warfare Cadre in the Falklands

The Royal Marines Arctic Warfare Cadre maintained forward observation posts from 21 May and throughout the conflict, providing vital information on enemy movements. In particular, they covered the planned routes of advance for British forces towards Teal Bay and Port Stanley. They were selected for this role due to their unique specialist training in cold-weather warfare in remote northern climates, such as the Hebrides and Norway. This enabled them to stay in position for long periods in the often-severe Falklands weather conditions. They carried with them enough supplies to survive for at least a week, along with radios and observation equipment and personal weapons.

The AW Cadre observed the insertion of observation posts by their opposite numbers in the Argentine forces, which would have provided the Argentine commanders with information about British advances. On 31 May, Captain Boswell of the AW Cadre and 19 men were inserted by an 846 Naval Air Squadron helicopter and attacked an Argentine observation post on Mount Simon. Other Argentine observers also surrendered, eliminating this threat to the advancing commando brigade.

Special Boat Service in the Falklands

On the night of 30–31 May, a six-man team of the Special Boat Service was inserted at Port Salvador, from where they carried out a reconnaissance of Teal Inlet. Their task was to report on the suitability of Teal Inlet for forces landing and advancing to Port Stanley.

CHAPTER 8

Military Scouting, the Global War on Terror and Beyond

From the Gulf War through the Global War on Terror that followed the 11 September 2001 terrorist attacks in the United States, military scouting and reconnaissance became more closely associated with the special operations community. Special reconnaissance was often carried out behind enemy lines and was often combined with direct action missions and target acquisition.

The Gulf War (August 1990–January 1991)

After the Iraqi regime under Saddam Hussein invaded and occupied Kuwait in August 1990, it was condemned by the UN Security Council, which sanctioned the removal of Iraqi forces from Kuwait by 'all necessary means'. The coalition arrayed against Iraq (about 700,000 personnel) consisted largely of US forces (540,000), along with considerable support from Saudi Arabia and other Arab nations including Egypt, as well as the United Kingdom.

On 16–17 February 1991, Operation *Desert Storm* began with a massive air campaign designed to neutralise Iraqi air defences, communications, command centres and other essential assets, including oil refineries and vital bridges. The era of smart bombs had arrived. Iraqi military forces were also attacked on the ground.

On 24 February, in Operation *Desert Sabre*, US-led coalition forces moved into Kuwait from Saudi Arabia. Kuwait City was soon relieved. Coalition forces also accomplished a massive left hook to attack Iraqi Republican Guard units that were held in reserve in Iraq.

By the time of the Gulf War, US reconnaissance units had adopted heavier armoured vehicles, moving from light tanks to main battle tanks and from Jeeps and armoured cars to fully tracked vehicles. For operations that still required a soft-skin general utility vehicle, the long-serving Jeep had been replaced by the High Mobility Multipurpose Wheeled Vehicle (HMMWV) or 'Humvee'. Tests showed that the Humvee could still be the preferred option where speed and agility was more important than armour. This would prove to be relevant in a particularly extended reconnaissance operation during the Gulf War known as the *Scud Hunt* (February 1991).

Special Forces in the Gulf War

Due to their rapid deployability, high levels of training, language skills and knowledge of the area, both US and British special forces were deployed to the Gulf region, in the British case as early as January 1991. On the US side, this included 5th Special Forces Group, 1st Special Forces Operational Detachment-Delta (SFOD-D), or Delta Force, SEAL Team Two and 19th Special Operations Aviation Unit. On the British side, almost the entire Special Air Service Regiment was deployed along with the Special Boat Service.

Special operations forces were placed all along the border of Saudi Arabia and Kuwait to carry out reconnaissance on enemy movements and installations.

Scud Hunt

As coalition forces gathered in Saudi Arabia and, as it became clear to Saddam Hussein that he would not get off lightly with his invasion of Kuwait in 1990, alongside the bluster, including threats of the 'mother of all battles', there was a particularly alarming development. The Iraqi regime possessed several Soviet-made MIM-104 surface-to-surface missiles which had been used in the Iraq-Iran war. These were now aimed at Tel Aviv and Jerusalem, as well as Saudi Arabia with civilians as the main target. The aim of Saddam Hussein was to break up the fragile coalition by provoking Israel into retaliation. He would then invoke Arab loyalty against a common enemy. Amid diplomatic efforts to restrain Israel, the US deployed Patriot anti-ballistic missile systems in Israel to shoot down the Scuds as they came in, while aerial reconnaissance attempted to identify the position of the Scud launchers after they were fired in order to destroy them.

The Soviet R-17 ballistic missile, known as SS-IC Scud-B in NATO terminology, had a range of up to 900 kilometres and were launched either from fixed sites or from mobile launchers on either MAZ-543 or Saab-Scania tractor trailers. In order to deceive their adversaries, the Iraqis deployed decoy vehicles and also concealed the launchers under bridges, highway underpasses, gullies or culverts. They also perfected the art of firing and moving the missile launcher back into cover within about 30 minutes, a practise known as 'shoot and scoot', giving very little time for coalition assets, such as F-15 or F-16L fighter bombers, to identify and acquire the target with their air-to-ground munitions.

UK and US Special Forces Reconnaissance in the Gulf War

The American Coalition Commander-in-Chief, Herbert Norman Schwarzkopf Jr, had a sceptical attitude to special forces and doubted their usefulness in an area as vast as the Iraqi desert. On the other hand, the British Army commander in the Gulf, General Sir Peter Edgar de la Cour de la Billière, had a background in the Special Air Service and had no doubts about the abilities and potential of special forces. British special forces had already been inserted behind enemy lines in Iraq by 20 January.

As the pressure grew to provide convincing solutions for the naturally aggrieved Israeli defence establishment and public opinion,

Schwarzkopf allowed himself to be persuaded that special forces could be employed in the campaign to seek and destroy the Scud launchers. As a result, from 7 February, US 1st Special Forces Operational Detachment-Delta (SFOD-D) were inserted into Iraq as part of a joint US-British mission to find the Scuds.

Delta Force was assigned an area above the east–west highway to Baghdad and the SAS the area to the south. Some of the British special forces were inserted by RAF Chinook helicopters. A and D Squadrons of the SAS were equipped with adapted Land Rover 110s, which included the necessary gun mounts for General Purpose Machine Guns (GPMGs), as well as a mount for a MILAN anti-tank missile system. Other vehicles included Unimog transporters and Armstrong trials bikes. B Squadron SAS were not so lucky. There were only the shorter-wheel-base Land Rover 90s available and these had no gun-mounts. Bravo Two Zero decided that the Land Rover 90 would be more of a menace than an asset and opted to deploy on foot, with disastrous results. Their radios failed when they called for assistance and the weather conditions were much more extreme than they had been advised. US Special Forces deployed with specially adapted special operations Humvees. The special forces were also tasked with finding and destroying fibre-optic cables that ran from Baghdad and provided essential input for the Scud launchers.

The SAS set up road watches reminiscent of the Long-Range Desert Group in North Africa during the Second World War. Those units that had vehicles would attack the Scud launchers with MILAN anti-tank missiles and they used the Minimi machine guns and grenade launchers to see off any Iraqi military retaliation. Delta Force's preferred weapon appears to have been .50 Calibre armour-piercing sniper rifles to disable the Scud missiles and eliminate its crews.

Although the combined US and UK special forces operation did not entirely close down the Scud threat, in the words of General Schwarzkopf to the British forces' commander, it had the effect of 'totally denying the central corridor of Western Iraq to Scud units'. Like the operations of Lawrence of Arabia and the Arab revolt, the presence of elusive but effective forces in the region had the effect of tying down large numbers of Iraqi forces which could not ascertain how many coalition forces were in the area. The deployment of elite forces in the Scud hunt also convinced the Israelis that everything possible was being done to neutralise the Scud threat.

UK 16 Air Assault Brigade – Pathfinder Platoon

The Pathfinder Platoon had its roots in the Second World War, when airborne units landed before major paratrooper drops to secure a drop zone (DZ), mark out the drop zone and set up equipment such as Eureka beacons and coloured smoke to guide the pilots. They were also trained to defend the DZ until the main drop was completed and pass back any relevant intelligence.

During Operation *Overlord* (6 June–25 August 1944) the 22nd Independent Parachute Company were tasked with several key missions, including an air assault on the Merville Battery on 6 June 1944, which they completed despite pilot navigational errors and consequent dispersal of units. The 21st Independent Parachute Company, preceding the 1st Airborne Brigade, flew in 12 Stirling bombers from Gloucestershire during Operation *Market Garden*. The company included several German and Austrian Jews who were especially eager to fight the soldiers of the Third Reich.

After disbandment at the end of the war, the Pathfinders re-emerged in various forms, including the No.1 (Guards) Independent Parachute Company, which served in several conflicts, including Cyprus (1956 and 1964), Amman Jordan (1958), Borneo (1964, 1965) and Northern Ireland.

The new Pathfinder Platoon was created in 1985 and became part of the UK official rapid reaction force, 16 Air Assault Brigade, in 1999. Personnel were drawn from all three parachute battalions and the Pathfinder Platoon was under the direct command of Brigade Headquarters.

Selection for the Pathfinders is rigorous and is modelled on special forces selection. Selection takes place in areas such as the

Brecon Beacons and the Black Mountains and includes contact drills, setting up observation posts (Ops) and an extended reconnaissance exercise.

As 16 Air Assault Brigade works closely with both the US 82nd Airborne Division and the French 11e Brigade Parachutiste, the Pathfinder Platoon regularly trains with the American and French counterparts.

The Pathfinder Platoon in Operation Palliser

The Pathfinders made a notable contribution to the UK's intervention in Sierra Leone in May 2000.

In July 1998, the United Nations Observer Mission in Sierra Leone (UNOMSIL) was established following a long period of fractured government in Sierra Leone as well as incursions by the Liberian Revolutionary United Front (RUF). The threat from the RUF continued to grow and they threatened to invade the capital, Freetown.

The UK Joint Rapid Reaction Force (JRRF) was despatched to the region by sea and air in order to hold essential facilities, including the airport, and allow foreign citizens to evacuate Sierra Leone.

The arrival of the 1st Parachute Regiment was preceded by the Pathfinder Platoon of 16 Air Assault Brigade. Their task was to carry out reconnaissance and to take, mark and hold suitable drop zones and helicopter landing areas.

On 17 May, RUF guerrillas made a night attack on the village of Lungi Loi, about 60 kilometres from Lungi Airport. Members of the Pathfinder Platoon had taken up defensive positions around the village and monitored the approach of the guerrillas with night-vision goggles (NVG). The Pathfinders opened fire on the rebels with their SA80 personal weapons and general-purpose machine guns (GPMGs), killing four of the RUF guerrillas. It was reported later that the soldiers had difficulty with the safety catches on some of the SA80 rifles, preventing them from firing. The Pathfinders were reinforced by C company of 1st Parachute Regiment, who were based at Lungi Airport, and there was also aerial support from Gazelle helicopters and Nigerian UN troops.

Operation *Enduring Freedom* (October 2001–December 2014)

Following the attacks by al-Qaeda in New York and Washington on 11 September 2001, which destroyed the twin towers of the World Trade Centre, damaged the Pentagon and caused almost 3,000 casualties, the United States and its NATO allies mounted an attack on the Taliban regime in Afghanistan which provided sanctuary for al-Qaeda and Osama bin Laden.

American and British special operations forces were some of the first to arrive in Afghanistan. Joint Special Operations Task Force-North included 5th Special Forces Group, 19th Special Forces Group, 2nd Battalion, 10th Special Operations Mountain Regiment, Joint Special Operations Task Force-South, including 3rd Special Forces Group and 3rd Battalion 160th Special Operations Aviation Regiment. In March 2002, special operations forces came under the command of Combined Joint Special Operations Task Force (CJSOTF).

Task Force Dagger

US Army special forces, or Green Berets, landed in the north of Afghanistan on 19 October 2001 and co-ordinated with a group called the Northern Alliance who were inimical to the Taliban regime. The soldiers carried out reconnaissance for enemy units and used targeting equipment to call in Allied air support in conjunction with Northern Alliance attacks on enemy ground forces.

The 12-man special forces units were inserted by C47-Chinook helicopters of 160th Special Operations Aviation Regiment (SOAR). The operations of Special Operations Forces (SOF) and the Northern Alliance culminated in the capture of Mazar-i-Sharif and the overthrow of Taliban forces in the north. In due course, Kandahar would be taken along with Taloqan and Kunduz.

Battle of Tora Bora (6–17 December 2001)

The overall reconnaissance mission objective for US, British, Australian, Canadian, German and other special forces during Operation *Enduring Freedom* was to track down the al-Qaeda leader Osama bin Laden. It was thought that he had retreated to the rugged cave complexes in the east of the country known as Tora Bora and, in due course, special operations forces, elite forces, such as the US Rangers, as well as other military units began to concentrate there. The Tora Bora caves provided storage facilities for substantial al-Qaeda stores and armaments.

US and Canadian forces wait to be extracted by a C-47 Chinook helicopter during Operation *Tora Bora*. (US Department of Defence)

Initially, special forces from 5th Special Forces co-ordinated a heavy aerial bombing campaign. These were supported by Delta Force, the British Special Boat Service and US 24th Special Tactics Squadron.

After a truce, which it is now thought may have enabled Osama bin Laden and his lieutenants to slip over the border to Pakistan, the fighting resumed. During this period both British Special Boat Service and German *Kommando Spezialkräfte* (Special Forces Command, KSK) special forces carried out reconnaissance in the area and provided support.

In retrospect, although special forces had provided reconnaissance and targeting information, the attack on Tora Bora was not followed through by elite allied forces who were based in the area, such as the US Rangers, but was left to local forces. There is a possibility that with a determined follow-through Osama bin Laden may have been captured.

Operation Anaconda *(1–18 March 2002)*

This operation took place in the Shah-i-Kot Valley of the eastern province of Paktia, and the plan was to provide an assault force (Task Force Hammer) and a cut-off force (Task Force Anvil) that would force the al-Qaeda fighters in the area to flee into a trap.

An Advance Force Operations Team (AFO) provided reconnaissance and intelligence about enemy locations and movements. The AFO was part of Task Force Bowie, which comprised Joint Interagency Task Force-Counterterrorism (JIATF-CT), the intelligence and reconnaissance elements of the special operations forces committed to Operation *Enduring Freedom* (OEF). The AFO deployed for Operation *Anaconda* consisted of 45 special operations forces operators from Delta Force, Naval Special Warfare Development Group (DEVGRU/Seal Team Six) as well as US Air Force tactical controllers from US Air Force 24th Special Tactics Squadron (STS). The AFO sent small teams into the valley to set up observation posts.

When the team callsign Mako 31 tried to move into position, it found that its planned observation post location was already occupied by al-Qaeda fighters armed with a 12.7mm DShK heavy machine gun, which could have severely compromised any force advancing up the valley. Mako 31 was spotted by the enemy and a short firefight ensued before the Mako 31 team called in an AC-130 Spectre gunship to eliminate the al-Qaeda position. They also requested that the AC-130 provide aerial intelligence on enemy positions and movements using its infra-red optics. However, the AC-130 experienced a navigational error in its systems and it engaged coalition forces advancing up the valley rather than al-Qaeda forces. This blue-on-blue incident was then exacerbated by mortar fire from the al-Qaeda positions. It soon became clear, from the weight of fire coming from enemy positions, that the number of al-Qaeda in the area had been seriously underestimated. Original estimates had been around 250 al-Qaeda fighters; the revised number was closer to 1,000.

In order to provide better reconnaissance for TF Hammer, it was decided to insert the other SEAL Team, call-sign Mako 30, on the mountain of Takur Ghar (3,191 metres), whose prodigious height gave it excellent potential for an observation post. After some confusion over timings, Mako 30, which included the US Navy SEALs and an Air Force Combat Controller, were flown up the mountainside in an MH47-E Chinook helicopter flown by 160th SOAR on 4 March. The helicopter's spec ops suite included adverse-weather capability, terrain-following radar, forward-looking infra-red (FLIR) and low altitude, high-speed flight for rapid infiltration. Against all standard operational rules, the SEAL commander chose to be inserted into the observation position itself rather than into an offset position which would have allowed them to approach the position covertly on foot.

As the helicopter approached the drop zone, it was hit by a rocket-propelled grenade (RPG) and by machine-gun fire which cut hydraulic cables. These spewed hydraulic fluid onto the deck inside the helicopter and one of the Navy SEALs slipped out of the back of the helicopter. He fell about three metres before hitting the ground. Despite being injured, he fired back at the al-Qaeda fighters. The pilot managed to move the helicopter out of the area before crashing 7 kilometres away.

A second helicopter was despatched to retrieve the Navy SEAL who had been left on the hill. This helicopter also came under fire, but it managed to drop the SEAL team and the Air Force combat controller before getting away.

The SEAL team moved back up the hill and engaged with the al-Qaeda fighters. During the firefight, the Air Force Combat Controller, Sergeant John Chapman, was hit, presumed killed, when attacking an al-Qaeda machine-gun position. One of the Navy SEALs was wounded. An AC-130 gunship overhead provided covering fire while the SEAL team withdrew.

A US Army Ranger Quick Reaction Force (QRF) was then despatched from their base at Gardez to the north. As their helicopter arrived over the location, it was also fired on by the enemy. A door gunner was killed and both pilots were wounded. The helicopter crashed and, as the Rangers left the wreck, two were killed. Another was killed while still on board. The remaining Rangers got themselves into a defensive position to take on the al-Qaeda fighters. A second helicopter carrying Rangers had landed further down the hill and this Ranger team now moved up the hill to support the Rangers at the top. Having regrouped, they launched an assault on the al-Qaeda positions and cleared them from the hill. The Rangers now came under fire from enemy positions on other hilltops nearby which pinned them down for the rest of the day, during which time another Ranger died. An Australian SASR observation team called in airstrikes to keep al-Qaeda reinforcements at bay. The Rangers were lifted out after dark.

Neither the SEAL team nor the Rangers were able to rescue the first Navy SEAL who had fallen from the helicopter. It was later discovered that he had been killed during a firefight with the enemy.

Conventional units continued to clear the Shah-i-Kot Valley until about 14 March, and AFOs then carried out reconnaissance in the Naka Valley to ascertain whether any al-Qaeda fighters were still in the area.

A convoy of al-Qaeda fighters that had assaulted the SEAL and Ranger teams on Takur Ghar was later ambushed by special forces and Rangers of Task Force 11. Once the al-Qaeda fighters had been defeated, articles of Ranger and 160th SOAR equipment lost on Takur Ghar were found in their possession.

Combined Joint Special Operation Task Force-South (CJSOTF-South) (Task Force K-Bar) October 2001–April 2002

Task Force K-Bar was led by Captain Thomas Harward of the US Navy SEALs and included a wide variety of both US and Coalition special operations elements, including US Army Green Berets from 1st Battalion 3rd Special Forces Group, US Navy SEAL Teams 2, 3 and 8; US Air Force Special Operations Command Combat Controllers and Pararescuemen; the Australian Special Air Service Regiment (SASR); Canadian Joint Task Force 2 (JTF2); Norwegian *Haerens Jegerkommando* (HJK), and *Marinejagerkommandoen* (MJK); German *Kommando Spezialkräfte* (KSK); Danish *Jaegerkorpset* and *Fromandskorpset*; and the New Zealand Special Air Service (NZSAS). Task Force K-Bar also had a close link with the US Marine Corps Expeditionary Units (Special Operations Capable).

By the end of the first month of their deployment in south-eastern Afghanistan, Task Force K-Bar had pushed hostile fighters out of the Kandahar area and were wresting adjacent areas from Taliban control. On 21 November, 20 US Navy SEALs were inserted into Objective Rhino as part of the staging for the introduction of US Marine Corps Task Force 58 into the area.

Overall, Task Force K-Bar's role was special reconnaissance. The various units from different nations brought their unique skills and expertise, including high levels of mountain training and experience by the Norwegian contingent.

Norwegian special operations forces carrying out reconnaissance during Task Force K-Bar. (US Navy SEALs)

The Canadian Joint Task Force 2 (JTF2) had the highest level of interoperability with US special operations units and was highly regarded for its professionalism. JTF2 carried out both special reconnaissance and direct-action missions during the period of their deployment.

Danish army and navy special forces carried out special reconnaissance during the Task Force K-Bar insertions and were also involved in the seizure of a Taliban commander on a joint operation with US Navy SEALs.

The German KSK were also assigned special reconnaissance missions with Task Force K-Bar and they carried out further missions with the NATO-led International Security Assistance Force (ISAF), which was deployed in Afghanistan between 29 December 2001 and 28 December 2014.

British Special Air Service (SAS) in Operation Determine

Both A and G Squadrons of the Special Air Service (SAS) were deployed to Afghanistan from October 2001. Their task was somewhat non-specific reconnaissance to track down al-Qaeda and Taliban forces or high-value targets. After a fruitless search in the rugged mountainous country, they returned to base in the UK.

In November 2001, C Squadron SAS was covertly inserted into Afghanistan to take part in a more focused operation alongside Task Force Sword to locate high-value targets including Osama bin Laden and his senior leadership team. A and G Squadrons were deployed to destroy an al-Qaeda opium plant.

British Special Air Service (SAS) in Operation Trent
November 2001

This operation involved a large deployment of SAS forces to attack an al-Qaeda opium plant in November 2001. So far as reconnaissance is concerned, Operation *Trent* broke most of the standard operating procedures (SOPs). There was no opportunity for preliminary ground reconnaissance before the operation was launched and it was carried out in broad daylight.

The operation began with a high-altitude low opening (HALO) descent by SAS Air Troop, who marked out the landing zone for the arrival of the main force. They were followed by six USAF C-130 Hercules aircraft which carried A and G Squadrons and 36 vehicles, including long-wheel base Land Rovers, or 'Pinkies' in SAS terminology, ACMAT VLRA liaison, reconnaissance and support vehicles, which carried supplies and reserves of ammunition, and scout trail motorcycles. As each Hercules landed, the ramp came down before the plane had halted and the vehicles sped off the back.

Once the entire assault force had disembarked and formed all-round defence, the scout trail bikes went off ahead to *reconnoitre* the route and the Land Rovers and other vehicles followed. Once within sight of the objective, the SAS troopers disembarked and, carried out a classic infantry assault. Heavy covering fire was provided

Special reconnaissance is facilitated by vehicles such as this Supacat HMT Extenda which has enhanced IED protection and manoeuvrability. (Supacat/SC-Group)

by MILAN anti-tank missiles, general purpose machine guns (GPMGs), MK-19 grenade launchers and snipers with L82A1 Barret 0.5in anti-materiel rifles with Leupold M series 10x magnification telescopic sights. Air support was provided by US Navy F18 Hornets, which destroyed the opium storage site.

Although the operation was successful, it may have raised questions as to whether the SAS was the appropriate unit to employ in what was, effectively, a standard infantry assault against a relatively low-value target.

British 3rd Commando Brigade

The British 3rd Commando Brigade was deployed to Afghanistan. In 2007, they were operating in Gamsir, in southern Helmand Province. A major Taliban stronghold was located in the area and it was decided to attack it in order to prove that NATO forces in the region could take and hold ground and to find useful intelligence and capture Taliban commanders.

The operation would go wrong for a number of reasons, but it is worth noting that both ground and aerial reconnaissance was carried out prior to the assault. Reconnaissance was carried out by both the Brigade Patrol Group, who were using Weapons Mounted Land Rovers (WMIKs), and by C Squadron Light Dragoons with Scimitar armoured reconnaissance vehicles. These moved close enough to the enemy to draw fire. They were able to establish that the fort was almost impregnable, with thick, high mud walls enclosing several compounds and that it was occupied. However, the crucial information that reconnaissance units were not able to ascertain was that the fort had hidden tunnels underneath it that could protect the fort's defenders during a bombardment. The tunnels were cleverly protected from thermal imaging cameras by scrub that was kept damp by the defenders. The tunnels connected the fort with compounds outside it, enabling them to shelter a large body of men.

Another obstacle between the commandos and the fort was the Helmand River. The commander wanted to ascertain where the river could be crossed by BvS 10 Viking amphibious all-terrain armoured vehicles prior to the assault. Reconnaissance revealed that the riverbanks were steep, but it was also important to try to identify a crossing point that was not within firing range of the fort's defenders.

The commanding officer of 1st Armoured Support Troop ordered a careful reconnaissance of the river before committing the Vikings to the river crossing. The reconnaissance, carried out by an expert in river reconnaissance, took two days but did not provide the CO with the answer he wanted. The recce report revealed that the river became less fordable as the distance increased from the fort and the CO was faced with crossing the river within 200 metres off the fort.

In the event, it was not the river crossing that would prove to be the problem during this assault. Once the Vikings had successfully crossed the river, Marines de-bussed for an infantry assault on the fort. However, by now the Taliban had emerged unscathed from their hiding places and poured unrelenting fire on to the advancing

Marines, who sought shelter in vain on either side of the Vikings. To make matters much worse, there was a fatal friendly fire incident.

British Parachute Regiment – Brigade Patrol Group in Afghanistan

The Brigade Patrol Group, part of the Brigade Reconnaissance Force, is responsible for long-range reconnaissance with the Royal Marine Commandos. An elite within an elite, it is made up of six four-man teams, each member of which has a particular specialism as patrol commander, signaller, medic and sniper All are trained parachutists and the group carries a range of weapons that may include the L96 sniper rifle, M16 assault rifle, L129A1 Sharpshooter rifle, Minimi light machine gun and a portable anti-tank weapon.

The Brigade Patrol Group has a particular specialism in cold and mountain warfare and some of their members are recruited from Mountain Leaders who have their roots in the Second World War

Pathfinders of the Parachute Regiment on a joint training exercise with French paratroopers. (Ministry of Defence)

Cliff Assault Unit. They are also trained in desert and jungle warfare and in covert surveillance and reconnaissance.

When the Brigade Patrol Group was based at Gereshik, in Helmand Province, the commander would send them out as combat recce patrols to scout the ground, especially the important heights which dominated the territory.

The idea was that the Brigade Patrol Group would be able to defend themselves if they were engaged by the enemy and/or call for support from the Light Dragoons who were on stand-by to assist if things got too lively.

Patrols would sometimes be mounted in vehicles or helicopters and sometimes on foot. The patrols would report on the lie of the land, likely areas for enemy contact, potential ambush sites and areas such as caves where enemy fighters might be concealed. Some operations also involved close target recces and directing strike operations.

The desert landscape was not ideal for covert recce patrols. The Brigade Patrol Group was, therefore, often employed for recce by force, whereby they would advance until there was some form of contact with the enemy, at which point they would report back the estimated size of the enemy force.

On another occasion, the Brigade Patrol Group commander was tasked with assessing an approach to a Taliban command and control post at Kashtay. The command post was to the west of the Helmand River and the Taliban were confident that the river was impassable. Attempts by the BPG to get a vehicle across indicated that the Taliban were correct in their assessment.

However, the BPG commander considered that they might be able to wade across, despite the fact that the river was in spate. To do so covertly at night would be particularly challenging, even more so when the ambient temperature was -6°C.

The BPG were transported to the river in Pinzgauer trucks, accompanied by Scimitar armoured vehicles. They then moved on foot to the crossing point near Kashtay. Four groups of six men tied themselves together before crossing in their underwear, carrying their clothing in waterproof bags. They then dressed once they reached the other side.

Soldiers were sent ahead to take out the sentries guarding the compound. The JTAC then made contact with a USAF Rockwell B-1B bomber to request a massive strike that would wipe out the compound and its occupants and enable the BPG to assault the compound to gather any useful intelligence.

In the event, the bombing was so devastating that the patrol commander considered it was futile to search for any useful intelligence in the wreckage and the patrol returned across the river, leaving Apache helicopters to finish off anything that was still standing.

Operation *Iraqi Freedom* (OIF) (March 2003– December 2011)

Operation *Iraqi Freedom* began on 20 March 2003, and achieved its objectives in about 21 days, namely the toppling of the Ba'athist regime of Saddam Hussein and the seizure of major cities such as Baghdad and Basra. It was characterised by speed of movement and by the successful coordination of coalition air, land and sea forces. Compared with the previous Gulf War, coalition forces covered ground four times faster.

Military objectives were achieved with the aid of accurate intelligence. However, the speed of movement placed considerable strain on the intelligence architecture. Information is not intelligence, and the constant flow of information from the perpetually moving front line had to be processed, taking into account various sources, before being used to make decisions for the troops on the ground. However, by the time this process had taken place, the situation on the ground had often evolved.

Aerial intelligence for coalition forces was gathered by about eight aircraft that included USAF Northrop Grumman E-8 Joint

Surveillance Target Attack Radar System (JSTARS); USAF Boeing RC-135 'Rivet Joint' Reconnaissance and signals aircraft; US Navy Lockheed P-3 Orion maritime patrol aircraft; RAF Hawker Siddeley Nimrod maritime patrol aircraft; USAF General Avionics MQ-1 Predator Unmanned Aerial Vehicles (UAVs); and USAF Northrop Grumman FQ-4 Global Hawk surveillance UAVs.

Special operations in Operation *Iraqi Freedom* were divided between task forces, some of which were focused on particular geographic areas.

Combined Joint Special Operations Task Force-North (CJSOTF-North) also known as Task Force Viking in Operation Iraqi Freedom

A plan to move conventional forces into northern Iraq was thwarted when Turkey withdrew permission. As a result, US 10th Special Forces assisted by 173rd Airborne Brigade were inserted via aircraft flying from Romania to conduct a special forces campaign allied with local Kurdish forces, not dissimilar to the operation in northern Afghanistan by 5th Special Forces Group. The objective was to keep Iraqi forces north of Baghdad engaged so that they could not move to reinforce defences in the south.

Combined Joint Special Operations Task Force-West (CJSOTF-West) 2003

CJSOTF-West was a coalition special forces operation in western Iraq that involved elements of the US 5th Special Forces Group, US Air Force Special Forces Command (AFSOC) Special Tactics teams; Australian Special Air Service Regiment (SASR) assisted by 4th Battalion (Commando) Royal Australian Regiment (4RAR); A and D Squadrons of the 22nd Special Air Service (SAS), C Squadron Special Boat Service (SBS) with support from 45 Commando Royal Marines, and signals and intelligence units.

The mission included reconnaissance and intelligence gathering, neutralising Scud launchers and striking at other designated targets.

Coalition Naval Special Operations in Operation Iraqi Freedom

A coalition Naval Task Force Group (NTG) was created for Operation *Iraqi Freedom* in 2003, which included US Navy SEAL teams, 8 and 9 Special Boat Teams, operated by Special Warfare Combatant Craft Crewmen (SWCCCs); US Air Force Special Tactics units; M Squadron Royal Marines Special Boat Service (SBS); 40 and 42 Commando Royal Marines; and Polish JW GROM special forces.

One of the primary assignments for naval special warfare units was to secure offshore oil and gas facilities to prevent them from being sabotaged by the Iraqi regime. These operations involved prior reconnaissance, typically by US Navy SEALs operating from SEAL Delivery Vehicles (SDVs), which can operate from submarines and enable covert insertion of reconnaissance and combat divers into the area of operations.

Naval Special forces did not only operate in the maritime environment, however. They also secured inland facilities including dams. This involved helicopter insertion and the use of a variety of desert special operations and reconnaissance vehicles, including the US Marines Desert Patrol Vehicle (LDV), designed to be carried in V-22 Osprey tilt-rotor aircraft.

Joint Special Operations Command Task Force in Operation Iraqi Freedom

This task force consisted of both US and British special operations forces as well as elite support forces, including the US 75th Ranger Regiment and the British Parachute Regiment. The special operations forces included US Navy SEALs, Delta Force, USAF 24th Special Forces Squadron, UK 22 Special Air Service, Special Boat Service, the Special Reconnaissance Regiment and 1st (UKSF) Signals Regiment.

During the conventional phase of operations against the Iraqi military establishment, the task force provided advanced special reconnaissance and target acquisition for coalition conventional forces, one example being the special reconnaissance carried out by Delta Force at the Haditha Dam where they identified the need for a much larger force to take the dam than had been originally envisaged.

As conventional operations merged into insurgency operations and the hunt for al-Qaeda leaders, the Joint Special Operations Task Force was given several different labels, including Task Force 21, Task Force 45, Task Force 88 and Task Force 714. Suffice to say that, despite the changing task force designators, this was a rolling operation and the mission remained largely the same. The different elements within the Joint Task Force organisation were broadly designated as Task Force Blue, including US Navy SEALs from SEAL Team 6; Task Force Green, including 1st Special Forces Operational Detachment (Delta); Task Force Black (later named Task Force Knight), including the Special Air Service and Special Boat Service; and Task Force Orange, including signals intelligence units.

Whatever the designation, the task forces ran a fast-paced and highly coordinated operation that involved special reconnaissance, surveillance, signals, computer, satellite and phone intelligence as well as sophisticated and highly nuanced interrogation techniques. Once information was gathered, special forces would strike decisively, and the intelligence wheel would continue to turn. Gradually, the intelligence teams worked their way closer to some of the biggest fish, including Abu Musab al-Zarqawi. He was responsible for a large part of the bombing campaign and beheadings in Iraq during the insurgency campaign. A visitor to his house was identified and tracked. On 7 June 2006 USAF F-16C jets dropped munitions on the house and Zarqawi was killed.

Such fast-moving operations by the task force resulted in over 3,000 insurgents being taken off the streets which led to a significant reduction in the bombing campaign.

UK Special Reconnaissance Regiment (SRR) in Operation Iraqi Freedom

The Special Reconnaissance Regiment is the newest component of UK Special Forces (UKSF). Formed in April 2005, it is designed to relieve the Special Air Service and Special Boat Service of a proportion of their reconnaissance, surveillance and intelligence gathering work in order to enable them to focus on direct action. It is also a product of the international response to the watershed terrorist attacks in the United States on 11 September 2001, with its implications for greater focus on intelligence gathering and vigilance. Operations *Enduring Freedom* and Operation *Iraqi Freedom*/Operation *Telic* also witnessed the importance of detailed reconnaissance and intelligence work.

As the largely successful conventional invasion of Iraq evolved into a more complex urban engagement with insurgents, the requirement for specialised reconnaissance and intelligence grew. In this world of shadows and whispers, covert intelligence in hostile urban environments could mean the difference between life and death.

US Marine Corps Reconnaissance in Operation Iraqi Freedom

During Operation *Iraqi Freedom*, the land and ongoing operations of the 1st US Marine Expeditionary Force (MEF) were preceded by Marine reconnaissance units. 1st MEF was first deployed to Kuwait, based on Camp Commando, before carrying out desert training before crossing the Iraqi border.

A Force Recon Battalion is assigned to each active Marine division and each Recon Battalion is divided into three companies. The 1st, 2nd and 3rd Reconnaissance Battalions are therefore attached to the 1st, 2nd and 3rd Marine Divisions, while the 4th Reconnaissance Battalion is the 4th Marine Division Marine Forces Reserve.

Marine Force Reconnaissance is characterised by deep penetration of the battlespace, where they operate inside the area of artillery support. The Marine Force Recon unit therefore needs to be able to defend itself if necessary, though on 'green' operations engagement with the enemy would not be its primary purpose. Their tasks may include one or more of the following: amphibious or ground reconnaissance to report on enemy activity and collect other useful information; surveying beach topography for amphibious landings or landing suitability for helicopters or parachute drop zones or aircraft forwarding operating sites; performing pathfinder duties in the relevant zones; conducting counter-reconnaissance against enemy units; and to conduct raids against designated targets.

Recon Marines can be inserted covertly by either water, parachute, including High Altitude Low Opening (HALO) or High-Altitude High Opening (HAHO), or by helicopter.

Information picked up on the ground by Marine Force Recon, such as updates on enemy movements, is relayed directly back to the Marine intelligence unit in the relevant Marine Expeditionary Force HQ.

The reports from a Marine Recon unit are valuable to a commander because they provide details from the ground in real time which would be difficult to obtain in the same detail from other means such as aerial photography.

The main difference between Force Recon and similar amphibious units, such as the Navy SEALs is that Force Recon's primary role is reconnaissance, whereas the SEALs, while they include reconnaissance as part of their training and mission, are primarily a strike force.

US Marine Corps 1st Reconnaissance Battalion in Operation Iraqi Freedom

The US Marine Corps 1st Reconnaissance Battalion conducts amphibious and ground reconnaissance.

During Operation *Iraqi Freedom*, the 1st Reconnaissance Battalion carried out patrols in Humvee vehicles. The battalion took part in Operation *Vigilant Sabre*, during which Marine scout snipers played a crucial role in identifying and eliminating insurgents in the city. The battle for Fallujah in November 2004 developed into a long-term struggle against insurgents which contrasted with the clear victory over Iraqi military forces in the early stages of Operation *Iraqi Freedom*.

Marine Forces Special Operations Command (MARSOC)

As special operations forces across all branches of the armed forces continued to grow in size and, as they became ever more frequently the weapon of choice of the Pentagon, the US Marine Corps initially retained full control of their special reconnaissance forces, such as Force Recon. This was because the Marines regarded themselves as an integrated force that should have central control of all aspects of their operations. However, it became apparent that Marines were not included in early operations where special operations forces were deployed under US Special Operations Command (USSOCOM).

As a first step to greater integration with the military special operations community, the US Marines created the Marine Expeditionary Unit Special Operations Capable (MEU SOC) in 2001, which was tasked with overt or clandestine direct action, recovery operations and special intelligence and reconnaissance operations. The emphasis was on targeted and short-duration operations. These operations were under the command of the theatre US Marines Corps commander and not US Special Operations Command.

As Operation *Iraqi Freedom* got underway, however, a Marine Corps Special Detachment 1, formed mainly from 1st and 2nd Reconnaissance Companies, was deployed as part of Naval Special Warfare Group One to carry out direct action and special reconnaissance. The success of this deployment contributed to the formation of Marine Forces Special Operations Command (MARSOC), which was created on 24 February 2006. This was a new command within the US Marine Corps while also a component of US Special Operations Command. As part of this process, some platoons from Marine Force Reconnaissance were assigned to MARSOC to form the nucleus of the Marine Special Operations Companies (MSOCs).

US Marine Corps Light Armoured Reconnaissance Battalions

Apart from its Force Recon and its special reconnaissance elements, the US Marine Corps also incorporates four Light Armoured Reconnaissance Battalions. Mounted in LAV-25 light-armoured vehicles, armed with a M242 25mm Bushmaster main gun, each vehicle can carry up to six Marine scouts in addition to the driver, gunner and vehicle commander.

The 1st Light Armoured Reconnaissance Battalion deployed in Operation *Desert Storm*, Operation *Enduring Freedom* and was first over the Kuwait/Iraq border during Operation *Iraqi Freedom* as well as one of the first US units to reach Baghdad.

The official brief for the LARB is 'reconnaissance, security and economy of force operations and … limited offensive and delaying operations that exploit the unit's mobility and firepower'.

US Marine Corps Scout Sniper

The logic behind the US Marine Corps Scout Sniper platoon is that the high levels of fieldcraft required of a sniper to get into position in order to eliminate a target are similar to those required for close scouting work. The telescopic sights used by both the sniper and the spotter provide a detailed view of enemy movements on the ground which is valuable for intelligence purposes.

A scout sniper platoon comprises between eight and ten scout sniper teams, with one member of each team designated as the sniper

A US Marine Corps Scout Sniper team. (Lance Corporal Sarah Anderson)

and the other as the spotter. The sniper may be equipped with either an M40 or an M82 sniper rifle, depending on the mission. If the mission includes disabling vehicles or firing through buildings, the 0.50 calibre M82 would be the weapon of choice. The sniper would normally be armed with an M9 9mm pistol for personal defence. The spotter would be equipped with night vision scopes and infrared laser equipment and armed with an M4 carbine.

The advantage of the scout sniper unit is that it can carry out both reconnaissance in detail and counter-sniping in urban areas where enemy snipers would be difficult to eliminate by aerial bombardment without causing collateral casualties.

US 75th Ranger Regiment: Regimental Reconnaissance Company

The Regimental Reconnaissance company was set up in 2005 to carry out special reconnaissance and target acquisition and is roughly on a par with the US Marine Corps Force Reconnaissance companies. It was formerly known as the Regimental Reconnaissance Detachment.

The Ranger Reconnaissance Company might operate in advance of the 75th Ranger Regiment battalions, carrying out Pathfinder missions, and it is also trained and equipped to operate and survive for periods of time without support from other units.

The Ranger Reconnaissance Company can be inserted by parachute (HALO or HAHO), aircraft or by boat.

US Army Reconnaissance, Surveillance and Target Acquisition (RSTA) units

These units consist of both mounted cavalry and unmounted infantry scouts. Some are trained as snipers, others in long-range reconnaissance, others specialise in communications and intelligence and others have responsibility for operating UAVs. There are also experts in fast-roping, combat swimming, small boat handling and air liaison.

Within the US Army, the reorganisation of Brigade Combat Teams reflects the growing appreciation of the importance of reconnaissance,

A British Army soldier prepares to launch a Desert Hawk UAV from a moving Land Rover. (Ministry of Defence)

whether it be infantry or armoured Stryker teams. Courses such as the US Army Reconnaissance and Surveillance Leaders Course, often attended by members of US special operations forces such as SEALs, or by elite forces such as the US Rangers, reflect the growing spread of special operations and elite forces organisational and operational principles throughout the regular army.

UK 1st Intelligence, Surveillance and Reconnaissance Brigade (1st ISR)

Made up of both regular and reserve units, the 1st ISR is responsible for acquiring information and delivering it for purposes such as target acquisition and intelligence assessments. It is also responsible for ISR and for operating UAVs. The reserve units include the Honourable Artillery Company (HAC) and both 21 and 23 Special Air Service Regiments.

A US Naval Construction Force 'Seabee' Reconnaissance Team (SERT) approach a bridge. (US Navy)

Surveillance and reconnaissance by reserve units such as the Honourable Artillery Company includes the establishment of covert observation posts to maintain 'road watch' style surveillance and to identify high value targets. Reconnaissance duties may be carried out by patrols. Specialist radar technology may be used to identify and counteract enemy mortar activity. Patrols that operate in forward areas or behind enemy lines are trained in escape and evasion (E&E) and resistance to interrogation (RtoI), as well as survival and patrol skills. Traditional and advanced signalling skills are also demanded.

Designed for low-altitude Intelligence, Surveillance and Reconnaissance (ISR), the RQ-11B Raven is a hand-launched and man-portable unmanned aircraft system 9UAS). The Raven can be manually controlled from the ground or set on a pre-arranged flight, lasting up to 90 minutes.

The RQ-11B Raven is one of the most widely used UASs, deployed to the US armed forces and the armed forces of many other countries.

Reorganisation of Military Reconnaissance Units

One of the features of the Global War on Terror was an increased appreciation of the value of special operations forces which led in turn to a rationalisation of reconnaissance duties and units.

The International Long-Range Reconnaissance Patrol School (ILRRPS) was established at Neuhausen ob Beck in 1979, and then moved to Weingarten in 1980. Initially, the training involved military personnel from Belgium, Germany and the United Kingdom and they were later joined by Greece, the United States, Norway, Italy, the Netherlands, Denmark and Turkey. Having provided the commanding officer for the school, the United Kingdom withdrew from the scheme in 2000 and the United States took over command. In 2001 the school was changed to the International Special Training Centre (ISTC).

The change of title for the school reflected the change of emphasis away from long-range patrol units and towards reorganised special forces units that included special reconnaissance in their skill set. The growing use of surveillance technology also gave military planners alternative options when gathering forward reconnaissance, though the eyes on the ground versus technology debate continues to run at the time of writing.

National reorganisations included the French *Commandement des Forces Speciales Terres* (COM FST), French Army Special Forces Command, which was established in 2002. Under this command was grouped the *1er Regiment de Parachutistes d'Infanterie de Marine*, with its roots in the Free French forces based in Britain during the Second World War and its connections with the SAS. This unit specialises in long-range reconnaissance and served in Africa, Bosnia and Operation *Desert Storm* in the Gulf. Another French special operations unit is the *13e Regiment de Dragons Parachutistes* (13th Dragoon Parachute Regiment). With a heritage stretching back to the Ancien Regime, the *13e RDP* special operations force is organised in eight squadrons, each of which has a particular regional expertise and operational focus across desert, equatorial and arctic environments and insertion by land, sea and air. The RDP took part in the Gulf War and used its reconnaissance skills to track down war criminals in Bosnia.

The German *Kommando Spezialkräfte* (KSK), Army Special Forces, and *Kommando Spezialkräfte Marine* (KSM) come under the overall organisation control of the *Einsatzführungskommando der Bundeswehr* (Joint Special Operations Command) The KSK was established in 1996 to increase German capabilities in special forces operations and it also absorbed the skill set of the previous long-range patrol reconnaissance company, the *Fernspahlerkompanie 200*, which was deactivated in 2011. Apart from its activities within Task Force K-Bar, the KSK also participated in Operation *Anaconda* when it contributed to the establishment of observation posts and reconnaissance missions to identify al-Qaeda escape routes into Pakistan.

Norwegian special operations combined the *Forsvarets Spesialkommando* (FSK) and the *Marinejagerkommandoen* (MJK). Apart from extensive operations with Task Force K-Bar, Norwegian special forces also operated in Kabul and Helmand Province, including special reconnaissance, hostage rescue and training.

The Danish *Specialoperationskommandoen* (SOKOM), Special Operations Command, was created in 2014 to bring the *Jaegerkorpset* (Jager Corps) and *Fromandkorpset* (Frogman Corps) under a single command. Both units had their commands transferred from the army and navy respectively.

Conclusion

In July 2016, the US Army announced that it would be deactivating its nine long-range surveillance (LRS) companies. In June 2019, the 1st Battalion 508th Parachute Infantry Regiment was the first US Army infantry battalion to deploy with the Black Hornet Personal Reconnaissance System, which they would take with them to Afghanistan.

These two events are capable of interpretation in several ways. On the one hand, the standing down of LRS units could be seen as a victory for technology over human reconnaissance. On the other hand, the significance of the highly portable Black Hornet nano drone may indicate that technology has enabled the democratisation of reconnaissance. Everybody can now become a military scout.

The trend can also be seen in the absorption of the duties of Long-Range Reconnaissance Patrols (LRRPs), such as the German FSLK200 (*Fernspählehrkompanie 200*), into special operations units. This does not mean that reconnaissance has been downgraded; rather that it has become more integrated. The logic is that special operations forces are often out beyond the front line and therefore are well placed for reconnaissance, even if their actual mission priority may be something else, such as direct action or target acquisition.

Elite commando units that bridge the gap between regular forces and special operations forces, whether amphibious or airborne, continue to value their specialised reconnaissance units, trained to go out in front, either covertly or to test the water with reconnaissance by force. Tough enough to manage on their own, they are also able to call in support either from the ground or the air.

Modular elements, such as the US Army Stryker Brigade Combat Team, or the new British Strike Brigades, included a reconnaissance element and dedicated vehicles. The US Stryker reconnaissance vehicle (RV) forms part of the Reconnaissance, Surveillance and Target Acquisition (RSTA) units. The latest British system was the Ajax armoured fighting vehicle which had a reconnaissance and strike variant to replace the Combat Vehicle Reconnaissance Tracked (CVR(T)) such as the Scimitar.

Whatever the form it takes, the principle of scouting and reconnaissance remains as important as ever.

Bibliography

Arrian, trs. De Selincourt, A. (1971), *The Campaigns of Alexander*, London: Penguin.

Baden-Powell, R. S. S. (1891), *Reconnaissance and Scouting*, London: William Clowes & Sons.

Burnham, F. R. (1927), *Scouting on Two Continents: A Life Spent Scouting in the West of America and South Africa*, New York: Doubleday.

Beevor, A. (2018), *Arnhem: The Battle for the Bridges, 1944*, London: Viking.

Field, R. (2003), *US Army Frontier Scouts, 1940–1921*, Oxford: Osprey.

Gebhardt, J. F. (2005), *Eyes Behind the Lines: US Army Long-Range Reconnaissance and Surveillance Units*, Fort Leavenworth, Kansas: Combat Studies Institute Press

Guderian, H. (1952), *Panzer Leader*, London: Michael Joseph.

Haythornthwaite, P. (2016), *British Light Infantry and Rifle Tactics of the Napoleonic Wars*, Oxford: Osprey.

Hesketh-Prichard, Major H. (1920), *Sniping in France: with notes on the Scientific Training of Scouts, Observers and Snipers*, London: E.P. Dutton and Co.

Herodotus, trs. De Selincourt, A. (2003), *The Histories*, London: Penguin Books.

Keegan, J. (2002), *Intelligence in War: Knowledge of the Enemy from Napoleon to al-Qaeda*, London: Pimlico.

McGrath, J. (2008), *Scouts Out!: The Development of Reconnaissance units in Modern Armies*, Fort Leavenworth, Kansas: Combat Studies Institute Press.

Neville, L. (2008), *Special Operations Forces in Afghanistan*, Oxford: Osprey.

Neville, L. (2008), *Special Operations Forces in Iraq*, Oxford: Osprey.

Plutarch, trs. Talbert, J. A. (1988), *On Sparta*, London: Penguin Books.

Rottman, G. L. (2012), *World War II US Cavalry Corps: European Theatre*, Oxford: Osprey.

Selous, F. C. (1896), *Sunshine and Storm in Rhodesia: Being a Narrative of Events in Matabeleland Both Before and During the Recent Native Insurrection*, London: Rowland Ward & Co.

Southby-Tailyour, E. (2008), *3 Commando Brigade: Helmand, Afghanistan*, London: Ebury Press.

Syrett, D. (2014), *The Eyes of the Desert Rats: British Long Range Reconnaissance Operations in the North African Desert 1940–43*, Warwick: Helion & Company.

Theotokis, G. (2019), *Twenty Battles that Shaped Medieval Europe*, Hungerford: Robert Hale.

Thompson, J. (1985), *No Picnic: 3 Commando Brigade in the South Atlantic*, London: Leo Coooper/Secker & Warburg.

Thompson, R. (1966), *Defeating Communist Insurgency*, New York, USA: Praeger.

Trench, C. C. (1985), *The Frontier Scouts*, London: Jonathan Cape.

Urban, M. (2010), *Task Force Black: The Explosive Story of the SAS and the Secret War in Iraq*, London: Little Brown.

Vegetius, F. R., trs. Clarke, J. (1967), *Military Institutions of the Roman (De re militari)*, Los Angeles: Enhanced Media Publishing.

Xenophon, trs. Warner, R. (1966), *A History of My Times (Hellenica)*, London: Penguin Books.

Xenophon, trs. Daykyns, H. G. (2019), *The Cavalry General*, Whitefish: Kessinger Publishing.

Index